Creation of a
Conscious Robot

JUNICHI TAKENO

Creation of a
Conscious Robot

Mirror Image Cognition and Self-Awareness

PAN STANFORD PUBLISHING

Published by

Pan Stanford Publishing Pte. Ltd.
Penthouse Level, Suntec Tower 3
8 Temasek Boulevard
Singapore 038988

Email: editorial@panstanford.com
Web: www.panstanford.com

British Library Cataloguing-in-Publication Data
A catalogue record for this book is available from the British Library.

Creation of a Conscious Robot: Mirror Image Cognition and Self-Awareness

ISBN 978-981-4364-49-2 (Hardcover)
ISBN 978-981-4364-50-8 (eBook)

Dedicated to the memory of my late father

Contents

Preface

This book introduces the author's current focus on the study of the consciousness and the mind. It also proposes a method for constructing the functions of consciousness and the mind on a machine. The author has been exploring the concept of a conscious robot for nearly 20 years. He has presented his ideas at two international conferences held in Florida, in the summer of 2003: The 7th World Multiconference on Systemics, Cybernetics and Informatics (SCI 2003) and the International Conference on Computer, Communication and Control Technologies (CCCT 2003). The paper presented at these conferences received the Best Paper Award in the system sessions at SCI 2003 and in the methodology sessions at CCCT 2003. The same paper again received the Best Paper Prize at the Innovations in Applied Artificial Intelligence section of the 18th International Conference on Industrial and Engineering Applications of Artificial Intelligence and Expert Systems (IEA/AIE 2005) held in Italy in the summer of 2006. His article titled "Mirror Image Cognition of a Robot" was honored by the academic society at the 3rd International Conference on Sensing Technology (ICST 2008), held in Taiwan. These peer acknowledgments of the merits of this study provided the impetus for writing this book.

This book is expected to be used as a textbook for university students pursuing information technology and mechanical engineering courses and is also suitable for students who are interested in the new areas of research. The content and information herein are also recommended for brain scientists and robot researchers, as well as those in arts who are enthusiastic about philosophical discussions and those who are curious about psychology and sociology. With such readers in mind, the author uses plain language to enable everyone to gain insights into the subject without needing expert knowledge. The initial sections of the book are dedicated to the discussions of basic knowledge about robots and the human brain. These sections intend to familiarize readers with the theme of this book. They are followed by an explanation of the conscious robot developed by the author. Last, experiments are introduced that demonstrate that this conscious robot — for the first time in the world — successfully recognizes its own reflection in a mirror. The author sincerely hopes that all readers interested in elucidating the mystery of human consciousness and the mind will benefit from reading this book.

Chapter 1

Introduction

One day, I was sitting in the phlebotomy room of a relatively large private university hospital, accompanying my elderly father. The room was rather quiet and several phelobotomists were not very busy. A boy, about second-grader in elementary school, was called by his name and number. He came to the phlebotomist, somewhat hesitatingly. The phlebotomist started the routine procedure, placing a band around the boy's arm and disinfecting the skin area. Just when she was ready to insert the needle, the boy said weakly, "I'm not mentally ready for this yet."

I and the other people around me were surprised by what the boy said. We were surprised because this boy, who was so young, used the words "not mentally ready." He clearly asserted the existence of his mind.

How was his mind formed?

In my everyday life, many ideas come to my mind and vanish before I know it. Not in a few cases, I concentrate on something and my consciousness is thoroughly absorbed in it. I sometimes glance at things absentmindedly. I might be impressed, may cry or may laugh when watching a movie. These activities are functions of humans and are known as human consciousness or mind.

Creation of a Conscious Robot: Mirror Image Cognition and Self-Awareness
Junichi Takeno
Copyright © 2013 Pan Stanford Publishing Pte. Ltd.
ISBN 978-981-4364-49-2 (Hardcover), 978-981-4364-50-8 (eBook)
www.panstanford.com

Why do humans have a consciousness or mind?

This question has been occupying people since ancient times.

The first few chapters of this book tell you the story of robots, including robots that are much talked about in the media today, and intelligent robots.

Then comes the story of the brain. A wide variety of research studies about the human brain, in particular that which is said to create human consciousness and mind, are introduced.

Another topic is consciousness and the mind as discussed in the fields of philosophy and psychology.

I will further introduce the study by Prof. V. Braitenberg, ex-director of the Max Planck Institute, Germany. With his robot, information gained through the sensory system, such as tactile sensations, is processed in its internal artificial neural networks, and the output is used to directly drive a motor system such as electric motors. His concept gave rise to the idea of building a scheme of a living creature, such as an insect, in a robot.

Another study introduced in this book is that by R. Brooks at the Massachusetts Institute of Technology in the United States. His innovative robot design ideas had a significant impact on researchers after him.

The discussion continues to artificial neural networks and evolutional technology. The neural networks used by Braitenberg are relatively simple, but the generally known and famous neural networks are discussed here. Readers learn that these neural networks assist in achieving the high-level motion of robots, including the recognition of objects.

A technique to build a scheme of the evolution of life in machines is explained. One of the advantages of this technique is that one can design artificial neural networks to transmit information input from the sensory system to the motor system relatively freely.

The subsequent chapters in this book are dedicated to introducing researchers attempting to create consciousness and mind in robots.

First, W. Grey Walter's robot is introduced. Walter developed the world's first biological-type robot.

Tadashi Kitamura developed a conscious robot. His robot features a hierarchical conscious structure and realizes a relatively high level of behavior such as anticipating the behavior of others.

Jun Tani also developed a conscious robot. Tani believes that consciousness occurs in the course of confusing cognition and prediction. He got this idea by observing humans. According to Tani, for example, when the usual way to school is suddenly blocked, one becomes conscious of oneself and asks, "What on earth has happened?" and becomes conscious of the surrounding external world at the same time.

Mitsuo Kawato also studies robots. He believes that the integration of the sensory and motor systems is essential to achieve complete human-like behavior in robots. This integration would enable robots to learn by imitation — one of the high-level behavioral functions of humans — and eventually acquire consciousness.

Kismet, a robot developed by Cynthia Breazeal at the Massachusetts Institute of Technology, has achieved fundamental communications with humans using the theory of the mind as discussed in the field of psychology.

After introducing these research studies, the book presents the conscious robot developed by the author. The design of artificial consciousness is described, followed by the design of the mind. The author's study is similar to those by Tani and Kawato, but the difference lies in that the methodology adopted by the author proposes a new paradigm regarding consciousness and the mind — one that encompasses and explains all of the questions that have been a mystery in the study of consciousness and the mind.

The author's methodology addresses the problems of not only learning by imitation but also imagination, qualia, stream of consciousness, and stream of feelings that appear as a representation of emotion and feelings. Qualia is a term that refers to how one "feels about something." For instance, just think of a cup of coffee, and you will "feel that sweet fragrance of coffee." According to some researchers, Qualia is one of the most difficult problems in the study of consciousness. In addition to qualia, the author's methodology is able to describe the human function of self-awareness (i.e., knowing that the image in the mirror is oneself), which is considered the supreme function of human consciousness. Understanding this function will be the first step toward understanding self-consciousness. The author holds that human consciousness has evolved and developed from the necessity of distinguishing oneself from others, and hence, his methodology describes the function for distinguishing between oneself and others.

The last chapter is dedicated to describing experiments on mirror image cognition using the conscious robot developed by the author and the results of those experiments. Three experiments were conducted: (1) The conscious robot imitates the behavior of the image of itself reflected in a mirror, (2) the conscious robot imitates the behavior of another similar robot, which is controlled via cables, and (3) two nearly identical conscious robots imitate each other. These experiments provide physical evidence that for the robot, the image of itself in the mirror is an existence that is closer to itself than part of itself is.

Finally, the author draws many suggestive observations from the experiments, including, for example, the possibility of developing artificial limbs that feel like the user's own limbs.

Let us now embark on an adventurous and intellectual journey!

Chapter 2

Story of Robots

When speaking of robots, readers may think of Sony's *AIBO* robot dog or Honda's *ASIMO* humanoid walking robot. Human-shaped robots are generally called humanoid robots. Both *AIBO* and *ASIMO* were developed by private businesses and are wonderfully animal-like and human-like, respectively, even from the viewpoints of robot researchers studying at universities. The creation and development of robots are introduced below.

As readers may already know, the word "robot" comes from Czech *robota*, literally meaning work or labor. It is commonly accepted that the English word "robot" was first used in a satirical drama *R.U.R.* (*Rossum's Universal Robots*, 1921), written by Czech playwright Karel Capek. In the drama, the robots created to work for humans eventually expel the humans. It is interesting to know that from the beginning robots were considered an existence to be watched with caution.

Later, robots were gradually exposed to the public, including an appearance in German movie *Metropolis* (1926), and in world expositions and various other exhibitions.

In the 1970 International Exposition in Tsukuba, Japan, many visitors were charmed by a robot playing the organ and a robot walking on two feet like a human. Before Honda surprised the world

Creation of a Conscious Robot: Mirror Image Cognition and Self-Awareness
Junichi Takeno
Copyright © 2013 Pan Stanford Publishing Pte. Ltd.
ISBN 978-981-4364-49-2 (Hardcover), 978-981-4364-50-8 (eBook)
www.panstanford.com

by revealing its *ASIMO* robot, there were many studies on robots that walked upright like humans.

People were surprised to see the *R2–D2* and *C-3PO* robots in American movie *Star Wars*. The cylindrically shaped *R2–D2* is a functional robot supporting fighter combat and navigation. The humanoid robot *C-3PO* walks on two feet, speaks human language, and even tells jokes. *C-3PO* is a protocol droid and understands the space language according to the setting.

Later, American movie *Terminator* featured a fearful and very tough robot, who performs impressively in a battle between the robots trying to expel humans and the defending humans.

I personally like movie *Andrew NDR114*, which is the name of a robot who fell in love with a human female because, for one thing, the movie raises a serious question about the "heart" of a robot.

Another memorable movie is *AI* (*Artificial Intelligence*). A robotic boy longs to become "real" and sets out to find a fairy. In the movie *iRobot*, a humanoid robot named *Sam*, who has a sense of self, saves the lives of humans and other robots. In the move *Wall-E*, a robot happens to acquire a heart one day and falls in love.

In Japan, *Tetsujin 28-go* ("Iron-man #28"), *Astro Boy* (Atom), and *Doraemon* (a manga series) are famous.

Astro Boy, in particular, is a typical humanoid robot. It is depicted as a hero of justice, fighting against evil together with humans. The physical ability of *Astro Boy* is tremendous. With visual power of 2.5 and hearing ability 1000 times that of a human, the robot can fly at Mach 5. To gain this speed, the robot is propelled by a jet engine in the atmosphere and by a rocket engine in space. It is said that when *Astro Boy* flies at Mach 5 in the atmosphere, the clenched fist of its arm extended in the direction of flight and its head would melt if ordinary metals were used because temperatures would reach 1500°C. At present, ceramics are known to withstand temperatures of 1500°C. On the other hand, *Astro Boy's* skin is said to be made of soft materials such as plastic. So far, there is no material suitable for *Astro Boy* that is both soft like human skin and highly thermoresistant. *Astro Boy* is basically a machine and does not have a heart. In the story, *Astro Boy* agonizes over its lack of a heart. I think this agony is a proof of the existence of its heart.

I have so far introduced various robots appearing in images, movies, and anime. These robot stories depict the interaction between humans and humanoid robots and at the same time the "light and shadow" of their interaction. Light represents the bright side of the view of a world in which humanoid robots cooperate with humans and help them realizing their dreams. Shadow represents the dark side and warns that humanoid robots could surpass and supersede humans.

Let us now go back further to the history of robots. In Europe, Heron (ca. 200 BC to 150 BC), a Greek inventor devised many automatically operating machines using weights in around the first century BC. And he invented a steam engine. Heron invented a device to automatically open the heavy doors of a temple when fire was used in a ceremony. This is an automaton rather than a robot, but autonomous robots may have their roots in automatons.

Clock technology saw remarkable development in Europe in the 18th century. Various mechanical dolls were built using clock technology. One of the exquisite achievements at that time is *The Writer*, an android made by Jacquet-Droz and his son. *The Writer*, a marvelous work, holds a quill with one hand, dips it in an inkwell and writes sentences on a sheet of white paper. These dolls are considered the roots of humanoid robots.

In 1738, the French Academy of Sciences in Paris introduced the automaton *Duck*. According to some reports, *Duck* did eat and defecate, but the truth is unknown. In the 17th century in Japan, the Takeda Karakuri troupe led by Oumi Takeda opened the *Karakuri* (mechanical doll) booth in Dotonbori, Osaka, and staged performances of the Takeda *Karakuri* dolls. According to historical records, the show attracted many people. *Karakuri* dolls are basically a machine or a doll remotely controlled by humans. They are the origin of present-day remote-controlled robots.

Well-known animated dolls include *Chahakobi* (tea serving doll) and *Yumihiki Doji* (arrow shooting doll) produced in the Edo period (1603–1868). The famous *Yumihiki* (Fig. 2.1), standing about 20 cm tall, holds an arrow in its right hand and a bow in the left. It slowly moves its right hand, holding the arrow, toward the left hand that holds the bow, places the arrow on the bowstring, and draws the bow. It turns its face to the target and positions himself to aim at the

Figure 2.1. Yumihiki Doji (arrow shooting doll) at the Toshiba Science Museum.

target. After fully pushing forward the bow by the left arm, the doll releases the arrow, and the arrow hits the target beautifully. The doll repeats this feat until it shoots all the arrows. An arrow rest holding the arrows rotates automatically a certain distance at a time for the doll to pick the next arrow always in the same position after each shot. It is surprising that the doll hits the bull's eye automatically. To reveal the secret, the doll releases the arrow always in the same direction and at the same speed so that you simply need to place the target in the position where the arrow will hit. The wonder of this animated doll is not that the arrows accurately hit the target but that such minute movements are automatically performed so smoothly without the slightest slipping.

A guidebook for constructing *karakuri* dolls entitled *Karakuri Zui* (*Mechanisms Compiled and Illustrated*) was published around 1796. This book may be the oldest robot engineering manual in Japan.

Traditional *Bunraku* puppetry is still appreciated by Japanese people today. It is performed by puppeteers who manipulate the puppets. These puppets are directly controlled by humans,

Figure 2.2. Unbelievably human-like robot (at the International Robot Exhibition).

but it may be said that they are mechanical dolls in a broader sense. *Bunraku* puppetry is good at expressing human feelings, or emotions, in puppets in an exquisite way that has never been seen before because the puppets are directly manipulated by humans. Now, the technique of robot in Japan seems very sophisticated (Fig. 2.2).

Japan is known to be a nation with great robotics expertise. As a proof of this statement, over 50% of all industrial robots working in plants throughout the world operate in Japan. Furthermore, 70% of the world's industrial robots are manufactured in Japan.

Industrial robots are quite different from *Astro Boy*, *ASIMO*, or any other humanoid robots. No humanoid robot has ever been used for industrial purposes. Humanoid robots are not yet good for practical use. There are many reasons for this. Two major reasons that are generally cited are (1) the effectiveness of bipedal walking is not yet established, and (2) the safety is uncertain.

Obviously, traveling on wheels is more efficient than walking as far as moving is concerned. Moreover, the robot's gait is unstable because a complex mechanism is used to achieve bipedal walking. If the robot falls over while working, human operators could be injured. The robot itself might also be damaged.

A humanoid that is capable of *ukemi* (literally, a quick response to any action) when falling down on the ground is reportedly to be developed soon. In the future, this problem of falling over will be solved (Fig. 2.3).

We now go on to a story about industrial robots. A floor-mounted arm mechanism called a manipulator is the basic form of an industrial robot. Since the manipulator is mounted on the floor, it cannot change its position or move. In a mass-production plant, products are transferred on assembly conveyors one after another.

Figure 2.3. HRP Robot (at the International Robot Exhibition).

Figure 2.4. Industrial robots at work (at the International Robot Exhibition).

Any mechanism designed to process products transferred on the assembly line does not need to move by itself. A manipulator is a mechanical structure that simulates the structure of a human arm, elbow, wrist and hand. So we see a number of artificial arms working at plants (Fig. 2.4).

The *Unimate* robot, developed in the United States in 1961, was the world's first practical industrial robot. As a typical industrial robot, *Unimate* affected the development of industrial robots thereafter.

The remarkable development of industrial robots is due to the working principle of "teaching playback." An operator performs a job step by step and the robot memorizes the entire procedure. After learning, the robot repeats what it has memorized by itself. Human operators perform the work just once to teach the robot. Thereafter, the robot plays back the steps repeating the job endlessly until the power is turned off. One might say that the robot performs a kind of "imitation learning."

Industrial robots perform spot welding, painting, and assembly more safely and better than humans. Spot welding is a technique for joining the metal of the car body not by welding linearly but in dotted lines. Linear welding provides stronger structures than spot-welding but requires higher welding accuracy. When welding structures, spot welding in the required locations by the required number of welds is better suited for robots.

Arc welding is a popular technique for welding metals. An arc is an electrical discharge like a spark. High voltage is generated on metal surfaces to produce sparks, and the resultant high temperature is used to fuse the metals.

Assembly robots are extensively used in the semiconductor industry for inserting electronic components into substrates. This job was done by humans in the past. The robot used for this purpose is generally called a SCALA robot (Selective Compliance Assembly Robot Arm). A SCALA robot is a simplified version of the manipulators already mentioned. A SCALA robot is called a horizontal articulated robot and consists of a vertical linear axis and one each rotating axis in two perpendicular directions. Its features include high positional accuracy in a three-dimensional space and the capability of moving parts from one position to another at high speed. These are the reasons why such robots are good at inserting devices.

A large number of manipulator and SCALA industrial robots have been installed in assembly plants throughout Japan to produce low-priced and high-quality products despite the disadvantage of variable-model short-run production, making Japan a nation with great robotics expertise and an economic giant.

Having introduced industrial robots, I will now proceed to the story of robots developed by researchers. Robots created by researchers are deeply related to the study of artificial intelligence (AI). Let us first review the study of AI.

The subjects in the AI studies curriculum at the Massachusetts Institute of Technology (MIT) Artificial Intelligence Laboratory in the United States include language, deduction, learning, vision, perception, operation, programming, architecture, and expert systems. Of these, vision, perception, deduction, programming, learning, and operation relate to the study of AI and robots. This suggests that the study of robots shares many common research areas with the study of AI.

Themes in AI studies include intelligent vision systems, perception function of robots, robot programming and AI, robot hands and tactile sensors, and autonomous mobile robots (Grimson and Eric, 1987).

The study themes for the intelligent vision system include technologies to represent image information captured by vision sensors as line drawings, detection of 3D shapes through the combination of lasers and vision sensors, and detection of 3D shapes by irradiating moiré interfering light and random dots in place of lasers. Another important theme is the development of an automatic system to represent an object viewed by vision as line drawings; decompose the line drawings into basic figures; and describe the relationships between the figures' connections.

The central theme in the study of the perception function of robots is stereovision. In the study of stereovision, at least two vision sensors set apart a certain distance are used to observe the object. The difference of how the object is viewed between the two images captured by the two sensors (generally called disparity) is calculated to determine the distance to the object by triangulation. This technique is simple in principle, but the problem is how to find the correspondence relationship between the images required for calculating disparity. The image correlation method is extensively used, but a decisive solution has not been yet established. The currently available solution is to increase the number of vision sensors for observation to enhance the reliability of the measurements. Most stereovision algorithms succeed in bounding the image domain to be used when searching for the correspondence relationship between images using an epipolar constraint. The epipolar constraint means that in stereovision, assuming a common plane passes through the two imaging planes of the two vision sensors and the object, the image information of the object present on the common plane is always reflected on the image planes of the two vision sensors where these image planes intersect the common plane and nowhere else. Other viable methods include optical flow to measure the movement of the sensors by tracking the images appearing on the sensors.

Robot programming and AI are described now.

The playback method is generally used when building and programming industrial robots. Programming a robot means that an operator does the job in the machine shop and the process is recorded in the memory of a robot for learning. With this system, the robot can flexibly respond to uncertain events and adapt to sudden changes in the environment such as an unexpected object moving in its area during operation. To respond to a variable environment, the robot needs to incorporate an environment observation system such as vision sensors and run a program capable of driving the manipulator flexibly.

Studies on sensitive adaptation to a variable environment belong to the domain of AI study and may be said to be equal to the study of intelligent robots. Many research papers were published in this particular field between 1970 and 1990. Some researchers' activities during this period are introduced here in detail.

Study themes in this field include planning problems, rule-based systems, model-based systems, and motion planning.

Early famous studies on planning problems include GPS (General Problem Solver) by Newell and Simon (1963) and STRIPS (Stanford Research Institute Problem Solver) by Nilsson et al. (1971). Using GPS, given the start state S and goal state G, the procedure from S to G is automatically generated. The monkey and banana problem is a famous toy problem in the study of artificial intelligence. Like GPS, STRIPS calculates the procedure from S to G automatically based on hierarchical planning.

Studies involving mobile robots include the famous SHAKEY developed (1966–1972) by the Stanford Research Institute (SRI) in the United States and the SHRDLU program at MIT that puts blocks together.

SHAKEY, the mobile robot, is connected to computers, has a TV camera and range finder (distance meter) serving as vision sensors. Driven by two electric motors, the robot moved around the room and was able to avoid obstacles of a simple shape. The behavior of the robot was automatically calculated by STRIPS. For example, for a robot in room1, which is given a goal state of in_room(room5) or moving to room 5, STRIPS writes a program reading go_through(door1, room1, room2), go_through(door2, room2, room4), and go_through(door3, room4, room5). Following the specified procedure, the robot executes each go_through instruction, which is a primitive action routines built into the robot, three times, and moves to the goal state.

This study marked a milestone in the history of robots as a successful installation of an automatic move procedure creation program in a mobile robot and one in which the robot actually moved as programmed.

As readers may have noticed, such procedural programs must be frequently re-written to achieve a different goal each time the environment changes. There is also a problem of combinatorial explosion in the hierarchical problem-solving method that STRIPS uses.

STRIPS and SHRDLU are examples of a rule-based system. When calculating work procedures from the start state S to the goal state G, the robot picks up currently viable work procedures using certain rules, and given the present state, the work procedure with the highest possibility of achieving the goal state is identified temporarily.

In a model-based system, all the information on the environment where the robot will have to work is loaded beforehand as models into the robot to save time and enable the robot to accurately calculate the environment using its vision sensors and other means.

Avoiding obstacles is one of the themes discussed in robot motion planning. What if, for example, SHAKEY enters a room and encounters an obstacle? This kind of problem is generally called a robot path planning problem.

Two solutions are available: a visibility graph and configuration space. A visibility graph is a graph in which all vertices and points of tangency of a curved section of obstacles visible from all vertices and points of tangency of any other obstacles are connected by straight lines (that is, if two vertices or points of tangency can see each other, a segment is drawn between them). The configuration space method provides a technique of increasing the size of obstacles relative to the size of the robot to be used. If the robot is a cylindrical one 30 cm in diameter, for example, it can be treated as a point by increasing the girth of an obstacle by, say, 15 cm. The robot then calculates the shortest path to the goal in the visibility graph while considering itself to be a point.

Studies on autonomous mobile robots are described now.

Autonomous mobile robots were first invented by William Grey Walter (1910–1977) in United States in 1948. The robots, named Elmer and Elsie, responded to light stimulation and recharged their batteries automatically. These robots will be further described in detail later in this book.

SHAKEY was certainly an intelligent mobile robot.

The CART project led by Hans Moravec at Stanford University is a well-known study on mobile robots (1980). Their robot moved on a four-wheel drive and communicated with computers by radio. Contrary to the SHAKEY robot, which was programmed with all the models of the moving environment, the CART robot had none of these models. Instead, the CART robot constructed the environmental models by itself while moving around in the environment by evaluating the uncertainty of its position using a nine-sensor stereovision system. After this, Moravec went to Carnegie Mellon University and created the NEPTUNE robot. This robot moved indoors at a speed of about 100 meters per hour while generating environmental models using ultrasound and stereovision sensors.

The author of this book developed an autonomous mobile robot equipped with a stereovision system consisting of two vision sensors in 1987. This was the world's first robot capable of detecting humans and other moving obstacles and foresaw and avoided collisions with moving obstacles using its vision system (Fig. 2.5) (Takeno, 2003).

Figure 2.5. Experiment of robot avoiding collision with a moving obstacle (Meiji University).

Martin Marietta Corp. developed ALV (Autonomous Land Vehicle) robot with the support of the U.S. Defense Advanced Research Projects Agency (DARPA). This robot used a small truck loaded with multiple computers, laser range finders, and vision sensors. It moved outdoors at a maximum speed of 10 km per hour while measuring the direction of its travel.

Large Jeep-like automobiles were then developed, incorporating the small-sized and high-speed computers, which became available in the market in and after around 1995. These vehicles were capable of traveling on rough terrain at nearly practical speed. Unmanned driving experiments are currently being conducted on highways and deserts in the United States.

Professor Ulrich Rembold at University of Karlsruhe, Germany, developed a robot named KAMRO (Karlsruhe Autonomous Mobile Robot) that featured two assembly manipulators and was capable of moving in all directions on a four-wheel drive during 1985 through 1994. The wheels had special driving mechanism to enable the robot to move in all directions as if it was sliding. The practicality of the robot deserves special mention: The robot was capable of assembling parts with two arms while moving around from one shop to another inside a plant. In October 1993, KAMRO successfully demonstrated self-controlled traveling and performed assembly jobs in a pseudo-experimental plant located on the campus of the university (Takeno and Rembold, 1996). KAMRO was equipped with 14 ultrasound sensors around its body and a stereovision sensor system to look ahead for obstacles as it traveled. The stereovision sensor system was developed by this book's author in Japan and mounted on KAMRO in Karlsruhe with Prof. Rembold (Fig. 2.6). The stereovision sensor system consists of a two-lens optical system, a vision sensor, and measured distance in 399 directions at a time (Takeno, *et al.*, 1992).

Later in Japan, Honda Motor Co. developed the Honda P Series humanoid bipedal walking robots. People all over the world were astonished to see the robot because its development had been done secretly. Until then, many robot researchers said that it was impossible to build robots that could walk on two feet like humans. After Honda's success, humanoid bipedal walking robots became the mainstream of such robots in Japan. Honda's *ASIMO* and Sony's *AIBO*

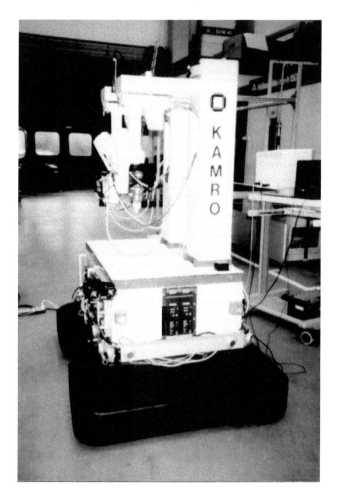

Figure 2.6. Dual-arm robot KAMRO developed by University of Karlsruhe.

robot dog became very popular. In the United States, the technology to drive unmanned car-shaped mobile vehicles was established and commercialized. In Europe, studies to understand the cognitive function of humans were actively being pursued although there were some other types of research being conducted in Germany such as those already mentioned in this chapter.

Chapter 3

Story of the Human Brain

This chapter provides some knowledge of the human brain to facilitate an understanding of human consciousness and the problem of the heart. I am not a brain science specialist; so please allow me a somewhat naive approach in my writing here.

The average human brain has a size of about 1500 cubic centimeters, and it contains 100 billion nerve cells called neurons and glial cells to protect the neurons. I had an opportunity to hold an actual human brain in my hands. It was heavier and smaller than I had expected.

Each nerve cell has dendrites as signal inputs and a fibrous neural pathway called an axon as the signal output. Signals pass through the axon as electrical impulses or transient changes in the potential. Each nerve cell has one axon. Each axon has many branches at its end. The axon is covered by many myelin sheaths to insulate the signals passing through the axon and accelerate the transmission speed of the signals.

The end of an axon makes contact with the inputs, the dendrites, of other nerve cells across a small gap. Although we say "contact," the nerve fibers are not directly connected. This gap, or a junction called a synapse, is about 20 to 25 nanometers wide. Signals are transferred across this gap by making use of special

Creation of a Conscious Robot: Mirror Image Cognition and Self-Awareness
Junichi Takeno
Copyright © 2013 Pan Stanford Publishing Pte. Ltd.
ISBN 978-981-4364-49-2 (Hardcover), 978-981-4364-50-8 (eBook)
www.panstanford.com

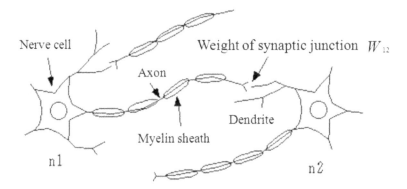

Figure 3.1. Outline of a nerve cell.

chemical substances called hormones. We may, therefore, say that all of the functions of humans are performed using two means of information transmission: electrical signals called spikes and chemical substances called hormones (Fig. 3.1). Specifically, the human brain transmits information to various parts of the body using two types of means: the pin-point type that mainly uses electrical signals to control the motion of the hands and legs, and the secretion type that injects hormones into the blood to transmit information throughout the body.

Each nerve cell has about 10,000 branches (synapses), that is, each cell connects to 10,000 other cells. The number of nerve cells in the human brain is fixed at birth. After birth, 100,000 cells die every year. It had been taken for granted that no new nerve cells are ever generated, but a recent report has refuted this long-held theory.

The brain looks like a walnut and consists of two hemispheres: right and left. It consists of a cerebrum, cerebellum, and the brain stem, which is a bundle of nerves that hang downward. The hemispheres are connected by a bundle of nerve fibers called the corpus callosum. The right hemisphere on the right hand side is called the right brain and the one on the opposite side is the left brain (Fig. 3.2a,b).

The cerebrum is the newest evolved part of the brain compared with the other older parts of the brain.

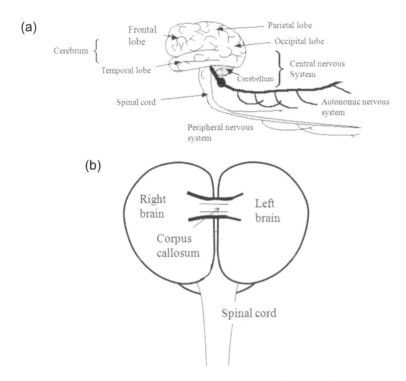

Figure 3.2. (a) Outline of cranial nerves. (b) Left brain, right brain and corpus callosum.

Earth was formed about 4.6 billion years ago. Scientists say that life first appeared on Earth 3.8 billion years ago. There were about 30 species of living creatures 3.6 billion years ago. About 800 million years ago, the number of species increased abruptly to 10,000. Living creatures in the sea moved to land about 700 million years ago. It was 650 million years ago when the big event of the extinction of dinosaurs occurred.

One theory holds that primitive human brains were formed about 600 million years ago and evolved to nearly the present form about 60,000 years ago. The cerebrum is responsible for high-level functions that make humans like humans, while the cerebellum is associated with animal functions such as motor control.

The surface of the cerebrum is covered by the cerebral cortex, which is divided into the neocortex and paleocortex. The neocortex

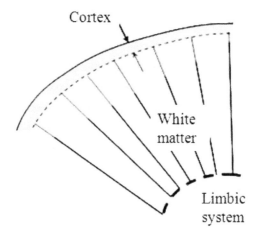

Figure 3.3.   Cortex and white matter.

is 2–3 mm thick and contains 14 billion neurons called columns in tiers. We can observe that the structure of the neocortex consists of about six layers. This area is supposed to be the home of the smallest functional units for processing terminal information to achieve the high-level functions peculiar to humans. According to scientists, the surface area of the neocortex is about 2,000 square centimeters, which is frequently referred to as equivalent to the size of a sheet of newspaper.

A number of nerve networks extend from the cerebral cortex down to the center of the brain. This area is called the white matter. The white matter consists of axons leading to nearly the center of the brain. The above-mentioned myelin sheaths glow in white (Fig. 3.3). Neurons and glial cells are the two main types of cells in the brain. Neurons transmit information while the glial cells supply nutrients to the neurons and remove unnecessary substances from them.

A small space called the lateral ventricle exists at the deepest part of the cerebrum. Its role is unknown. The limbic system extending around the brain stem below the white matter is an older part of the brain. The part responsible for controlling the irises of the human eyes is said to be located in the brain stem. A simple way for a doctor to test the functioning of a subject's brain is to shine the light of a flashlight into the eye of the subject and check the reaction of the iris.

The complete absence of any iris reaction indicates that the oldest and centermost part of the brain is not functioning.

The function of the brain is described now in brief. The right brain controls the nerve networks present on the left side of the body, performs artistic processing regarding meaning, emotion, and so on. The left brain, on the other hand, has its nerve networks mostly on the right side of the body; it handles languages and theoretical processing.

If the right brain is impaired, the left side of the body is paralyzed. Upon looking at a dog, a patient may be able to represent it as "dog" in speech but the semantic meaning of dog is thought to be lost. If the left brain is injured, the right side of the body is paralyzed and language abilities are disturbed.

The corpus callosum is a bundle of about 200 million nerve fibers.

The right and left brains exchange information constantly and both artistic and logical activities are carried out simultaneously. Upon looking at a red rose, one might represent the flower as "This is a rose" in speech and at the same time feel a sense of beauty aroused by the red color of the rose.

When the corpus callosum is severed, the right and left brains are nearly disconnected, and results in a critical condition depending on individual case. This type of surgical operation was once performed on epileptics. Normally no major problems occur, but troublesome cases of alien hand syndrome have been reported. Simply put, alien hand syndrome means being out of control of one's own hand. For instance, if one wants to write with the right hand, but the left hand tries to stop it. The person is conscious of writing with the right hand but the left hand moves without the person being conscious of it. Based on this fact, many researchers assert that the part governing human consciousness resides inside the left brain.

It has been said that by using some clever devices in an experiment, one can see that the part governing consciousness does exist in the right brain, but the relevant process will not surface and remains at an unconscious level. Some researchers thus believe, referring to the alien hand syndrome, that two parts of consciousness would arise if you severe the corpus callosum. Subconsciousness means consciousness that does not surface and thus fails to become conscious. Some call this unconsciousness, but unconsciousness normally refers to a loss of consciousness following anesthesia; so I differentiate how these two words are used in this book.

We can draw three general conclusions from the above discussions:

(1) The part governing consciousness resides in the left brain.
(2) There is just one consciousness that surfaces.
(3) There are two types of consciousness: the consciousness that surfaces (called explicit consciousness or simply consciousness) and subconsciousness.

The limbic system consists of the epiphysis, hippocampus, amygdala, thalamus, hypothalamus, putamen, and caudate nucleus. The epiphysis exists singly, whereas the others exist in right and left pairs. One of the important functions of the limbic system is described here. Based on a variety of information received from the sensory organs, the limbic system processes emotional information by communicating with the cerebral cortex, and it stores the information temporarily. At the same time, the limbic system generates a long-term memory by communicating with the cerebral cortex. Researchers say that the control mechanism for emotion and short-term memory exists in the limbic system.

Multiple nerve systems are interconnected with the limbic system, building a complex mutual relationship among them. I recommend referring to the adequate references noted for details on the limbic system.

A simple description is presented here. The amygdala is closely related to human emotions. The amygdala sends impulses to the hypothalamus by communicating with the frontal lobe of the cerebral cortex and modulates blood pressure and heart rate using hormones. The change occurring in the body is returned to the frontal lobe of the cerebral cortex via the somatic perception area. It seems that this circulation of information contributes to the generation of emotions. It is also said that the amygdala stores memories of negative emotions such as fear.

The thalamus relays signals input to the brain, to various parts of the brain. Signals entering the brain first reach the thalamus. The information is then sent to the cerebral cortex and the signals are also sent to the amygdala. At this time, if the amygdala is allowed to function with priority, the person is strongly affected by emotion.

The hippocampus is responsible for memory. If the hippocampus is damaged, short-term memory is totally lost.

The putamen controls unintended or subconscious actions. It issues commands to the premotor area of the cerebral cortex, and the premotor area sends commands to the neighboring motor cortex. This results in muscles contracting to perform motor skills.

The caudate nucleus is said to control subconscious actions and thoughts.

The hypothalamus controls metabolic processes and desires.

The cerebral cortex is described here. The cerebral cortex is generally said to encompass the frontal lobe, parietal lobe, occipital lobe, and temporal lobe (Fig. 3.2a).

The frontal lobe is associated with the highest-level functions of humans. It is deeply concerned with the generation of thought, planning, and conception as well as emotions. The parietal lobe is highly associated with some types of motion, calculation, and cognition. The occipital lobe is known to be involved in vision.

The temporal lobe in the left brain is related to speech, and, in particular, Wernicke's area, which is located there, is famous as the part of the brain associated with the cognition of spoken and written language.

The angular gyrus, located immediately posterior to Wernicke's area, is very much associated with information on meaning (Fig. 3.4).

According to experiments carried out by Wilder Graves Penfield (1891–1976), the part of the brain involved in long-term memory spreads over a wide area of the temporal lobe.

It is known that sexual arousal is generated when the temporal lobe is stimulated, and the frontal lobe is active during both male and female orgasms.

The cingulate gyrus of the frontal lobe is where researchers believe self-consciousness resides. Self-consciousness is a function by which an individual is capable of judging that an image in a mirror is his/her own or someone else's. Self-consciousness is also known to be highly associated with Broca's area, which is responsible for speech functions. This suggests that being able to speak about

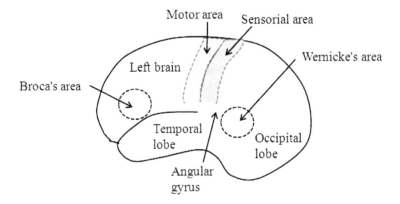

Figure 3.4. Broca's area and Wernicke's area.

oneself could be an origin of self-consciousness. Broca's area is located in the frontal lobe. It has also been said that the area involved in **will** exists in the prefrontal area of the frontal lobe.

The sensorial area and motor area exist in the region where the parietal lobe and the occipital lobe meet. The sensorial area is where the sub-areas corresponding to the tactile sensors of various parts of the body exist. These sub-areas are cortices and receive signals from respective tactile sensors. The cortices receiving signals from the hands, fingers, lips, genitals, and toes occupy a relatively large surface area (Fig. 3.4).

The sense of taste is associated with the cortex on the lower part of the right frontal lobe.

The stimuli of smell directly enter the limbic system and activate the lower part of the right frontal lobe.

The primary motor cortex lies at the center of the parietal lobe. It sends signals to activate the motor functions of the body. The right hemisphere's primary motor cortex controls muscles on the left side of the body, while that on the left controls muscles on the right side of the body.

When a human wants to make a certain movement, the primary motor cortex sends signals to contract muscles. At this point, there may be information circulating between the vision and somatic sensors in the body and the brain to monitor and make sure that the desirable motion is achieved.

The area involved in spatial cognition exists in the rear part of the parietal lobe. Most of the occipital lobe is used for vision processing.

An image entering through the lenses of the eyes is inverted when reaching the retinas. There are two types of photoreceptor cells in the retina: cones and rods. The cone cells detect intensity and the rod cells color information. The fovea in the retina is where photoreceptors are most densely concentrated. Signals from the photoreceptors, already image-processed on the retina to some extent, leave the eyes and go to the brain via the part called the blind spot and through a bundle of neural pathways. The blind spot consists of a dense bundle of neural pathways and there are no photoreceptors. The blind spot is therefore "invisible," or cannot respond to light stimulation. It is very interesting to note that the photoreceptors are viewing light that has penetrated the layers of nerve fiber bundles (from the back side of the nerve cables) since the photoreceptors are located at the deepest part or bottom of the retina (Fig. 3.5).

The neural pathways leave the right and left eyes and go to the brain. On the way, they cross each other just once at a certain location. This crossing point is called the optic chiasm.

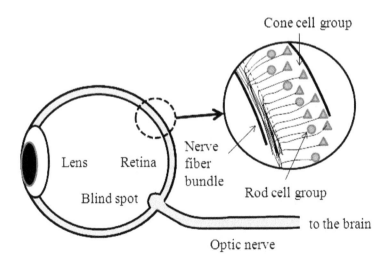

Figure 3.5. Vision and nervous system.

Because of this crossing, the neural pathways associated with the right-side visual fields of both the right and left eyes go to the left occipital lobe and the neural pathways associated with the left-side visual fields of both eyes go to the right occipital lobe.

Optic chiasm has not been fully explained biologically. The author supports the hypothesis that the optic chiasm has an advantage in that even when if one hemisphere of the brain is damaged, the other hemisphere of the brain keeps the functions of stereovision viable.

Visual information runs through the part of the brain called the lateral geniculate body, which is a part of the thalamus and reaches an area named the primary visual cortex (V1) in the right and left occipital lobes.

V1 is like a projector that shows image information captured by the eyes. Visual information is further transferred to other visual areas called V2, V3, V4, and V5 for the brain to determine the position of an object being observed. V2 processes three-dimensional perception, V3 distance, and V5 motion information (Fig. 3.6). The visual information passing through V1, V2, and V4 determines what the object is. V4 is responsible for processing color information.

When the primary visual cortex of the occipital lobe has a lesion, humans cannot see their outside world. It has been reported, however, that even with the visual area damaged, part of the visual

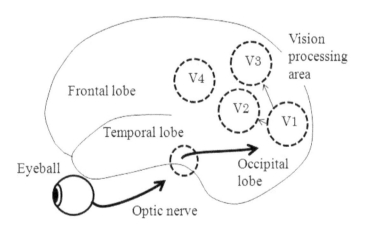

Figure 3.6. Vision processing system.

information was transmitted to the parietal lobe, and by virtue of this information the subject reached out his hands toward a strong light although he could not visually see the light. This is called blindsight or cortical blindness. In this case, the subject was not conscious of the light but reached out his hands toward it.

Chapter 4

Human Consciousness and the Mind

This chapter introduces historical research into the mystery of thought, consciousness, and the human mind, and, in particular, areas of mathematics, psychology, and philosophy. I believe an intricate mechanism of consciousness and mind exists in the background of human thought.

Hippocrates (460 BC–375 BC), an ancient Greek physician, was the first scholar who believed that human thought occurred in the brain (Bear, 2001). He said that the brain controlled the body. He reportedly discovered this fact through his experience in treating gladiators who suffered brain damage in ancient Rome.

4.1 Human Thought and the Turing Machine

Alan Turing (1912–1954) (Fig. 4.1) was the first person in history who mathematically challenged the mystery of human thought. The Turing machine he proposed in 1950 may be considered the origin of thinking machines. Turing said, "What can be solved by humans can be solved by a machine of a small capacity" (Turing, 1936). He thought that human intelligence could be realized by machines.

Creation of a Conscious Robot: Mirror Image Cognition and Self-Awareness
Junichi Takeno
Copyright © 2013 Pan Stanford Publishing Pte. Ltd.
ISBN 978-981-4364-49-2 (Hardcover), 978-981-4364-50-8 (eBook)
www.panstanford.com

Alan Turing
1912–1954

Figure 4.1. Alan Turing.

The Turing machine has a very simple construction. It consists of a body, a tape, and an input/output device called head to read and write symbols on the tape. The Turing machine heralded the development of present-day computers. In fact, it has been mathematically proven that all problems that can be solved with present-day computers can be solved with a Turing machine.

It should be remembered, however, that to create computers of today, we needed a stored-program system, which was developed on the basis of the ideas of John von Neumann (1903–1957) and program-specific high-level concepts such as algorithms and data structures.

It is also known presently that there is a problem that a Turing machine cannot solve. The problem is for a Turing machine to decide if it will finish running or will run forever. Any given problem is solved when the Turing machine finishes running.

Turing's attempt to create models of human thought paved the way to the development of computers, on the one hand, and the study of artificial intelligence, on the other.

4.2 Language and Formal Logic

We will start with a description of the study of formal logic before discussing mathematical ways to model human thought. According to formal logic, human thought arises from languages, and is an inference in languages. Formal logic was the basis of the study of languages and inference at that time. The development of modern formal logic is credited to Gottlob Frege (1848–1925), (Frege, 1892) and Bertrand Russell (1872–1970), (Russell, 1959). I believe that their objective of research was to unravel the intelligent functions of humans from the standpoint of formal logic. Their attempts, however, deviated from and became irrelevant to human thought and shifted to a pure pursuit of formal logic itself. They attempted to develop a method to derive mathematical knowledge from basic mathematical hypotheses using logic. Their efforts were in vain, however, when in the 1930s, Czech-born Kurt Gödel (1906–1978) mathematically proved that such an attempt was impossible (Fig. 4.2). These are Gödel's famous incompleteness theorems (*Gödelsche Unvollständigkeitssatz*). The theorems say that formal systems are incomplete because they inherently include arithmetic equations that are defined to be true by the system but that cannot be proved by the basic assumptions of the system. These theorems thus negated the clarity of mathematics. The meaning of these mathematical theorems also applies to the relationships between human thought and the computer as a representative of formal systems. Namely, there are some arithmetic equations that human thought believes to be true but that cannot be proved by computers.

Based on this recognition, Nagel and Newman (1958) held that computers could never reach the level of human thought or the human mind. In 1994, Roger Penrose asserted that human thought could clarify mathematical truths better than computers (Penrose,

Kurt Gödel
1906–1978

Figure 4.2. Kurt Gödel.

1994). According to Penrose, Fermat's last theorem was successfully proved by human thought, whereas if the Turing machine were used to solve that problem, it would have required infinite computations about whether or not the problem was solved to find the solution. He concluded that human thought solved problems that Turing machines could not solve, that is, that human thought involves non-computational elements that computers lack (Penrose, 1989, 1994). He referred to the quantum theory to explain the non-computational elements, but there are not many supporters of his argument today.

4.3 Language and Chomsky

Noam Chomsky (1928–   ) is a famous linguistic researcher who attempts to explain the philosophical aspects of humanity through the study of language. His arguments are introduced here. Chomsky's conclusions are based on the fact that humans speak languages and exhibit intellectual functions by conversing with others. Importantly, the study of artificial intelligence started with the study of language.

Chomsky thought that creating a machine that understands and speaks natural languages was a viable means to understanding the human mind scientifically. In his Theory of Generative Grammar, Chomsky showed that certain basic rules exist in the languages spoken by humans (Chomsky, 1957). This theory provided a scientific means to analyze human languages and contributed to the later development of the PASCAL and LISP computer programming languages.

Chomsky's theories have influenced many researchers. One of his famous studies involved chimpanzees that learned a "language" and were able to communicate with humans. Pictograms or signs (similar to a sign language) were used because chimpanzees do not speak human language; so a sort of surrogate language was used. The experiments were successful to some extent. The chimpanzees understood the surrogate language as well as the feelings shown by the humans. The experimenters also felt that the chimpanzees were willing to show their feelings to the humans using the surrogate language. The grammar was simple, however, and complex sentences were not used.

Chomsky's linguistic theory could not answer the question of why children can quickly learn a language as they grow up. This question is called Plato's problem. To find the solution to this problem, linguists started to study how language was generated and understood in the language areas of the human brain.

Austrian-born Ludwig Wittgenstein (1889–1951) was a famous philosopher who attempted to clarify human consciousness by analyzing the natural languages of humans (Wittgenstein, 1953). He tried to account for human conscious activities using the Language Game concept (Wittgenstein and Solipsism).

Jean Piaget (1896–1980), a Swiss developmentalist, was a famous figure in the study of how the functions of cognition develop in infants (Piaget, 1952).

As discussed in the previous chapter, it is very difficult to research the brain functions of humans. Human rights must be respected when treating the human brain. Furthermore, even though various systems have been developed to measure brain functions noninvasively (that is, without physical injury) it is still impossible to measure the true functions of the brain.

A functional magnetic resonance imaging (fMRI) system allows a doctor to view how the brain is activated in real time by measuring changes in blood flow in the brain. The system identifies and shows the active parts of the brain but does not provide any information on what the brain is performing at that part.

A transcranial magnetic stimulation (TMS) system is used to magnetically stimulate the brain cortexes through the skull and measure the reactions to examine the brain functions. This also does not reveal any information about the functions of the individual areas of the brain and of the brain as a whole.

Today's computers understand the natural languages of humans to some extent. Some computers are capable of generating languages similar to natural languages.

Chomsky's linguistic theories greatly contributed to the development of programming languages and artificial languages for computers. His theories also achieved their original aim in the research phase of natural language. Computers may understand programming languages and execute some processes, or even understand natural languages and follow instructions, but there remains the question whether computers understand languages in the same way as humans do.

Some researchers, known as behaviorists, believe that machines do understand. Behaviorists do not care what processes are going on within the machine, but they determine that the machine does understand just from the fact that the machine reacts adequately at least to certain stimuli. The soundness of this concept was questioned in a thought experiment conducted by John Searle (1932–).

This is a story of a box performing questions and answers in Chinese (Searle, 1980). The story is based on Searle's Chinese Room thought experiment, which is slightly modified. An Englishman, A, who does not understand Chinese at all sits inside a box. A has a complete manual to reply in Chinese to any question written in Chinese. The scenario of the thought experiment goes like this:

A Chinese person, B, writes a question in Chinese on a slip and puts it in the slot of the box. A takes it and consults the manual to find the reply written in Chinese. He copies the reply on a slip and gives it to B. B thinks that A fully understands Chinese because a perfect answer is given in Chinese to every question asked. However, the truth is that the person in the box is an Englishman who does not understand Chinese at all. If B writes a question in English, the person in the box will naturally understand the question and write the answer in English. The questioner would then think that the person in the box understands English, and his thinking would be right.

What is the difference between these two stories?

This question gave rise to the problem of human consciousness and the mind that exists beyond the problem of thought. This thought experiment makes us reconsider what human understanding really is.

4.4 Mind–Body Dualism of Descartes

French philosopher René Descartes (1596–1650) was the first Western philosopher who addressed the problem of the human mind (Fig. 4.3). With the famous declaration, "I think, therefore I am" (*Cogito, ergo sum* in Latin), Descartes insisted on the existence of the self. He stated, "I doubt sensation, logic and thinking, and after doubting everything, what is left is the fact that I think." This statement marks the discovery of the existence of the self of humans, or eventually, self-consciousness. Descartes advocated the famous mind–body dualism. The term appears in his work, *The Discourse on the Method* (1637) as *material–mind dualism*.

Mind–body dualism says, in essence, that the human being is divided into a mind and a body. Descartes thought that body exists in this world as a material substance, while the mind is connected to God. This thinking was favorably accepted in Christian Western Europe as a teaching that did not contradict Christianity. His thinking supported the Christian doctrine that stated that when a human dies, the soul (mind) leaves the body and enters the kingdom of God.

At the time, some people felt that clocks had minds as they watched them, but Descartes' mind–body dualism shattered this idea and convinced people that clocks were simply machines, and there was no mind in them. His thinking helped expel the subjective

René Descartes
1596–1650

Figure 4.3. René Descartes.

existence of the mind from machines and continued to develop further as the concepts supported modern capitalism in Western Europe, which enthusiastically pursued the rationality of machines themselves, now freed from the problem of the mind. It may not be an exaggeration to say that the Industrial Revolution that started around the end of the 18th century was backed by the concept of mind–body dualism.

Let us now look at the history of the study of consciousness and the mind. In 1879, the study of consciousness and the mind was started by Wilhelm Wundt (1832–1920), University of Leipzig, Germany, as part of psychology. Wundt explored his own psychological state using self-observation methods in which he studied consciousness and the mind by describing or speaking of his own mental state (Wundt, 1899). A while later, the self-observation method was discarded after it was criticized for its lack of objective reproducibility and condemned to be far from a scientific technique.

Next comes the Concept of Mind by Gilbert Ryle (1900–1976). According to Ryle, the Cartesian mind–body relationship, as understood by Ryle, was the dogma of the ghost in the machine; the Theory of Dual Existence of the privately observable life (the mind) and the publicly observable life (behavior) is a basic category mistake (Ryle, 1949). A category mistake refers to discussing two matters at the same level while ignoring the categorical difference between them, such as behavior and the mind at the same level.

Ryle states, "Mental episodes are analyzed as **disposition** to behavior." This statement became the first step in expelling the myths of ghosts in machines. However, this statement, at the same time, encouraged another myth called behaviorism, which will be discussed in detail later in this book.

Ryle's assertion gave birth to the mind–brain identity theory, which became popular in the 1960s as advocated by U.T. Place and J.J.C. Smart in Australia.

The mind–brain identity theory holds that mental processes are identical to cerebral processes.

From the 1960s to the 1970s, D.M. Armstrong and D. Lewis disseminated the mind–brain identity theory. Armstrong emphasized that the physical state of the central nervous system determines behavior. Lewis preached that sensation plays a causal role in the generation of sensation by certain stimuli, and beliefs play a causal role in certain behaviors (Ryle, 1949).

H. Putnam further developed the mind–brain identity theory. In 1967, he established a theory in which the mental state was viewed as being identical to the state in a Turing machine. This was the beginning of functionalism. Functionalists view the mental state as being identical to the machine state (Putnam, 1999).

Both behaviorism and functionalism were destined, sooner or later, to face the problem of qualia, which is difficult to solve. The qualia problem will be discussed in detail later.

4.5 Behaviorism and Cognitism

Russian physiologist Ivan Pavlov (1849–1936) (Fig. 4.4) succeeded in his experiments on conditioning the psychical reflex using the

Ivan Pavlov
1849–1936

Figure 4.4.   Russian physiologist Ivan Pavlov (1849–1936).

salivary glands of dogs. Pavlov won the Nobel Prize for this discovery (Pavlov, 1927).

Pavlov's experiments showed for the first time that conditioned stimuli generated mental reflexes. His experiments led to the birth of a new philosophy called behaviorism.

Behaviorism takes the standpoint that consciousness and the mind are mental representations of reactions to stimuli given to the body. Behaviorists do not discuss consciousness or the mind itself but believe that the functions of consciousness and the mind can be described as stimuli given to the body and the resultant reactions. It is true that behaviorism was considered more scientific than the self-observation method in that, compared with the impalpable notion of consciousness and the mind, stimuli and reactions could be discussed and shown to people in a reproducible way.

Behaviorism was established as a theory by U.S. psychologist John B. Watson (1878–1958), who was inspired by Pavlov's experiments. According to Watson, the study of consciousness itself cannot be a subject of scientific investigation. His thinking remains overwhelmingly influential even today (Watson, 1930).

Behaviorism attempts to understand consciousness and the mind by observing stimuli and reactions. Later, behaviorists realized that spiritual activities involved some active aspects such as will and intention, and the new behaviorism was introduced to explain voluntary actions such as cognition and intention. The new behaviorism, however, departed from the original behaviorism concepts and gradually lost its centripetal force.

Cognitism emerged just at this time. Cognitism is an idea that the intelligence of organisms derives from an internal cognitive system. Cognitists started to re-consider the existence of consciousness and the mind and attempted to incorporate systems of perception, judgment, memory, learning, language, inference and problem solving into computers.

In the early days of the development of cognitism, however, the body, will, and feelings were not taken into consideration. In the beginning, cognitism devoted itself to the study methods of intellectualism and treated representation symbolically. Intellectualism pursues an objective and intelligent way of thinking and ignores subjective concepts such as feelings and will as much as possible. Representation means, in brief, the images of external objects that emerge on consciousness based on perception.

The aforementioned Turing machine and Chomsky's linguistic theory are preeminent achievements in the study of human consciousness and the mind in this era.

It became essential for researchers to directly address and elucidate the problem of consciousness and the mind in order to construct consciousness and the mind in computers based on functionalism.

Cognitive psychology founded by Donald Broadbent (1926–1993) was the first scientific approach directly targeted at studying consciousness. In his work *Perception and Communication* (Broadbent, 1958), Broadbent says that consciousness is similar to a communication channel, and the number of objects that consciousness can pay attention to at the same time is limited.

Telephony developed rapidly during Broadbent's lifetime. Broadbent thought that consciousness might be scientifically explained using an analogy of telephone lines. As he understood it, consciousness was the function of attention. He believed that consciousness could be scientifically described by focusing on this human function of attention. Broadbent was interested in knowing how humans were attentive to the information they received from their environment via bodily sensors such as eyes and ears. He assumed that human attention was restricted, or that the capacity was limited. He relied on two methods: split-span and shadowing methods.

Gray and Wedderburn conducted the following experiments in 1960: A mixture of numbers and words were input into the right and left ears of a subject. For instance, "8, Eat, 5" were input into the right ear, and "Mice, 2, Cheese" into the left ear. According to the Broadbent theory, the subject was expected to answer, "8, Eat, 5, Mice, 2, Cheese." But instead the subject replied, "Mice, eat, cheese, 8, 2, 5." The experiment revealed that when listening to human voices attentively, humans selectively pick up information input into the right and left ears. The subject selectively picked up information and constructed a piece of meaningful information, or a sentence, "Mice eat cheese," from the mixed information input into both ears (Gray and Wedderburn, 1960). This, of course, suggests that humans have a system to selectively utilize consciousness.

I would like to introduce those philosophers who attempted to look into the essence of consciousness and the mind. There are two main trends in modern philosophy: phenomenology and hermeneutics.

4.6 Phenomenology and Hermeneutics

The leading exponents of phenomenology and hermeneutics are Edmund Husserl and Martin Heidegger, respectively.

Germany-born philosopher Edmund Husserl (1859–1938) was an important historical figure who addressed the problem of

Edmund Husserl
1859–1938

Figure 4.5. Edmund Husserl.

consciousness directly as a theme of philosophy (Fig. 4.5). The philosophy he advocated is called phenomenology (Husserl, 1900).

Husserl tried to solve the problem of consciousness and the mind, which are separated from each other in Descartes' body–mind dualism. The most difficult point in dealing with consciousness and the mind is the fact that both are subjective matters. It is impossible to use objective techniques, that is, direct scientific techniques, to study consciousness and the mind. By developing phenomenology, Husserl introduced a technique to solve the problem of subject and object, using consciousness (a subjective existence) as the starting point. This technique was the same as the one used by scientists who promoted science since the Middle Ages, including Nicolaus Copernicus (1473–1543) and Galileo Galilei (1564–1642). Copernicus and Galileo conceived the hypothesis that the earth is not at the center of the celestial motion, but instead the sun is, that is, a heliocentric theory rather than geocentric theory. This hypothesis can be said to be the subjective thinking of Copernicus and Galileo. They started verifying the hypothesis by conducting experiments and celestial observation using astronomical telescopes. After repeating experiments many times, they arrived at a stage at that the hypothesis seemed true to anyone who might think about it, irrespective of time and place. This is the stage where a new conviction was realized in which the subject and the object were consistent with each other. This type of new conviction is called a paradigm.

A scientific technique is a method to describe knowledge from the viewpoint of exploiting the power of nature with its maximum energy at any given time, in any place and by anyone. This is the very technique employed by scientific positivism that supported the development of modern science.

In other words, Husserl's phenomenology addressed the core problems of modern scientific positivism.

Kitamura clearly describes what phenomenology is from the perspective of an engineer as follows:

> Phenomenology is a method that uses subjective analyses to tell the essential character of consciousness and works in the process of arriving at a conviction about something. Two basic functions of consciousness discovered through phenomenological analysis are intuition and orientation. Intuition is the ability to know the properties of an external object via perception, while ignoring relevant hearsay or information on the object. The orientation of consciousness refers to directing consciousness toward an object to define its meaning, and reflecting and elaborating on the above process in an effort to find a more accurate meaning. Phenomenology uses the methods of phenomenological reduction and intersubjective reduction to extract the essence of consciousness. Phenomenologists believe that by using these methods, a purely subjective attitude is possible. To put it simply, phenomenological reduction means to put aside all of our common sense about material objects, customary ways of thinking and convictions about the world — a state called epoché. By so doing, one can understand the fundamental way consciousness exists, that is, the way consciousness exists when one recognizes, feels, views, thinks and recalls.

Husserl also showed through intersubjective reduction that the cognition of others is based on a projection of one's recognition

(Kitamura, 2000). This reduction method of Husserl to get to the core of the problem of consciousness is famous. Husserl also believed that the essence of consciousness is a process in which the object of consciousness (*noema*) and the action of consciousness (*noesis*) appear alternately and eventually arrive at a meaning.

Kitamura considers that this process represents an engineering **feedback**. He established a machine consciousness theory, which will be discussed later in this book.

Scientific objectivity was highly esteemed in Husserl's time. However, it had a critical deficiency. When scrutinized in detail, alleged scientific objectivity is basically associated with a scientist's daily research activities, which are entirely supported by the subjectivity of the scientist. This means that discoveries of objective facts by scientists are based on the subjective views of those scientists.

The technique adopted by Husserl was to analyze how the consciousness of the self exists in humans using the reflection function inherent in the subjectivity of the self, and eventually to elucidate the universal function commonly existing in human consciousness. Readers may wonder if this technique is good enough to discover objective facts. It should be noted that even when scientific devices are used to take measurements, subjective judgments are inevitably involved in interpreting the measured data. This means that uncertainties in subjective judgment are not much different from uncertainties in so-called objective judgment.

As such, Husserl's phenomenology bridges the wide gap between the subject and the object. According to Husserl, although consciousness is a subjective matter, researchers can come close to the truth by circulating the objects of consciousness and acts of consciousness, and consciousness itself can be an object of consciousness. We may therefore say that Husserl was the first philosopher who established a scientific research method for studying the subjective matters of consciousness and the mind.

Martin Heidegger (1889–1976) was a German philosopher. "In der Welt Sein (*Being in the World*)" and other Heideggerean concepts gave rise to the great trends of cognitive science (Heidegger, 1962). Heidegger re-defined phenomenology as hermeneutics.

Hermeneutics is a methodology used to view and understand all things arising from the human mind as representations of experience, and attempts to grasp the human mind scientifically through interpretation called "understanding." *Sein und Zeit (Being and Time)* is one of the most important works of Heidegger.

Based on Husserl's phenomenological methodology, Heidegger analyzed the fundamental problem of human existence as an existence associated with the world as space and time [*In der Welt Sein (Being in the World)*].

The difference between Heidegger and Husserl lies in that Husserl analyzed consciousness and the mind as an internal existence in humans, whereas Heidegger interpreted consciousness and the mind as "being in the world." In other words, Heidegger believed that consciousness and the mind should not be considered an internal problem of humans but should be analyzed as they relate to the outside world, or environment and situations. The ideas proposed by Heidegger are now remembered as the starting point of contemporary cognitive science.

Heidegger's assertions became the source of the notion of "cognition embedded in situations" and "embodied mind" in the newly conceived cognitive science.

Embodiment is a hot topic of discussion in cognitive science today.

4.7 Phenomenology of Embodiment

French philosopher Maurice Merleau-Ponty (1908–1961) was the first person to explain embodiment philosophically. He started studying perception and embodiment based on the phenomenology of Heidegger and existentialism of Jean-Paul Sartre (1905–1980). His book *Phenomenology of Perception* is famous.

Merleau-Ponty denied the mind–body dualism of Descartes and his followers, declaring, "My consciousness is experienced in and through my body." He also said, "The mind is the cause of the body, and the body determines the condition of the mind."

French philosopher Sartre explains human existence using two opposing ideas: being-in-itself and being-for-itself. The former referred to the essence of substances and the latter to the essence of human consciousness.

According to Sartre, two different concepts coexist in humans. A human is itself a substance (being-in-itself) and at the same time differentiates between "this" and "not this" (being-for-itself) and endeavors to become an existence exceeding the substance itself, that is, becoming free. He asserts that freedom is the essence of human consciousness. Sartre stayed with the mind–body dualism concept, but, interestingly, he explained that freedom is the essence of human consciousness.

The thinking of Merleau-Ponty was succeeded by Emmanuel Lévinas (1906–1995), who became an influential thinker of the era. The difference between Merleau-Ponty and Lévinas is that Merleau-Ponty tried to explain **the other** based on **the self**, whereas Lévinas attempted to explain **the self** from the viewpoint of **the other**.

Materialism was promoted by German thinker **Karl Marx** (1818–1883) as an analogy of Hegel's philosophy (Fig. 4.6). Marx replaced Hegel's mind with material and the development of the

Karl Marx
1818–1883

Figure 4.6.   Karl Marx.

Sigmund Freud
1856–1939

Figure 4.7.   Sigmund Freud.

mind with the development of an economic society to establish historical materialism.

Austrian psychopathologist Sigmund Freud (1856–1939) (Fig. 4.7) shed light on the psychological principle existing in the subconscious of humans in his book, *The Interpretation of Dreams* (Freud, 1900). He thought that hysteria arose from sexual inhibition and found that hysteria was improved by treatments aimed at sexual release. Later, he generalized this theory and preached that many psychological diseases occurred from sexual inhibitions. It may be said that Freud showed in his book that the problem of the human mind was not yet solved in the era where historical materialism was at its peak.

Carl G. Jung (1875–1961) followed and developed Freud's thinking (Wehr, 1989). Jung believed that there were origins, something like the Great **Mother** and the World **Mandala** of Esoteric Buddhism, in the depths of the human mind, and tried to explain humans' mental world as emotions arising from these origins.

Norbert Wiener
1894–1964

Figure 4.8. Norbert Wiener.

Another noteworthy research of consciousness and the brain is cybernetics, a bio-machine theory. Cybernetics was defined by Norbert Wiener (1894–1964) (Fig. 4.8) as the science of communication and control in organisms and machines. Cybernetics covers a broad field of studies, including medicine, engineering, physics, biology, mathematics, psychology, and social science (Wiener, 1956). It was later divided into information science, bionics, cognitive science and robotics. The term cybernetics is formed from a Greek word that means *steersman*.

I have so far introduced historical researchers related to consciousness and the mind. I will now summarize the knowledge about consciousness and the mind itself.

One of the important problems regarding consciousness and the mind is, which approach should be taken: materialism or idealism?

The Cartesian mind–body dualism holds that the body and the mind are separate from each other, and the mind is connected to God. From the Cartesianist point of view, the mind would then be discussed as a matter of idealism, and the body would perhaps be explained from the viewpoint of materialism. It should be noted, however, that Descartes does not explain how the separated mind and body are interrelated.

Descartes' allegation seems to directly apply to present-day computers. Computer hardware is a materialistic existence, whereas software and data are idealistic. It may sound somewhat rude and unreasonable to say that the software and data stored in a computer are an "idea," but this might be permitted perhaps as an analogy. Software is information and never belongs to the category of material. Software can only have meaning when it exists side by side with a physical substance or hardware.

The central theme of this book is to discuss the problem of creating consciousness artificially; so I would like to consider consciousness and the mind materialistically. I discovered, however, that conventional materialism is insufficient to explain consciousness. I believe a new materialism should be conceived. This will be discussed later in this book.

The next question is, which should be used to explain consciousness and the mind: functionalism or behaviorism?

Functionalism is "a position in epistemology and methodology not to treat things as real matters statically but to consider their functions dynamically, as having correlations, and as a process" (the *Kohjien*, "Wide garden of words," a single-volume Japanese dictionary). Functionalism is, thus, a theory to consider consciousness and the mind as a set of a variety of mental functions.

Behaviorism is "a thought aiming at becoming an objective science of psychology. The target of psychology is limited to learning the lawful relationships between stimuli and externally observable reactions (behaviors), while eliminating concepts related to consciousness" (*Kohjien*; partly modified). According to this definition, "concepts related to consciousness are eliminated."

This book is intended to create artificial consciousness, and therefore I cannot accept behaviorism that specifically eliminates consciousness.

Behaviorism was the product of an age when scientism exerted an overwhelming influence upon psychology to completely remove subjective matters from psychology in an attempt to conduct purely objective and scientific study.

According to Husserlian phenomenology, everything starts from the subject. It is, thus, forever impossible for behaviorism to unravel the mystery of consciousness and the mind because it excludes the study of subjective matters.

In my opinion, the subjective approach of studying consciousness and the mind adopted by phenomenology will gradually shift to an objective approach. This gradual shift will probably take place when, for example, phenomenological hypotheses are scientifically proved as a result of advancements in physiological knowledge about the brain. It is also possible, on the contrary, that multiple scientific discoveries may stimulate the creation of a new phenomenological model for unifying them.

To start a study of consciousness and the mind from a phenomenological perspective, we must first establish subjective hypotheses about consciousness and the mind. This would require a functionalistic approach of some form or another. This is because it is easier to study consciousness and the mind by assuming that consciousness and the mind comprise a set of various functions at least in the initial stage of study.

If we side with the functionalists, we would face a problem in that various functions are individualized and reduced to their ultimate details. For example, the cerebral function will eventually be reduced down to the input and output state of stimuli at every brain cell. Accordingly, instead of simply relying on functionalism, we need to prepare a scheme for unifying the various functions.

The next discussion is about the relationships between the brain, on the one hand, and consciousness and the mind, on the other.

In Chapter 3, "Story of the Human Brain," I mentioned that human consciousness is not lost even when the right brain is damaged. From this fact, brain scientists believe that consciousness is closely related to the left brain.

When one engages in some action intentionally and attentively, the frontal lobe of the brain exhibits vigorous activity. If will and attention area function of consciousness, this would seem to indicate that consciousness is closely related to the frontal lobe. If the important function of the mind is the generation of feelings, the mind may be closely related to the limbic system of the brain.

Spiritual experiences, religious experiences, and subconscious activities are said to be related to the right brain. If subconscious is considered to be consciousness of which the person is not aware, consciousness would be closely related to the right brain (Sperry, 1974).

We cannot ignore the sensory sensors of humans because stimuli input from the environment via these sensors affect consciousness and the mind. It is also obvious that the movements of the muscular system are associated with consciousness, even though they might be unconscious movements. For these reasons, the author believes that consciousness and the mind are closely related to not only the brain but also the body.

The next question is whether the brain has representation.

As discussed above, representation refers to the images of external objects that emerge in consciousness based on perception. There are perceptual representations (the object exists in reality), memory representations (reproduced from memory) and imaginary representations (creating a mental picture). Representation is sensuous and concrete, which makes it "different from concepts and ideas" (*Kohjien*). I will explain this briefly. If you see an apple in front of you, the image of the apple is formed on the retinas of your eyes. The image stimulates the photoreceptors of the retinas, and the relevant signals are transmitted to the brain, first reaching an area called V1. The signals are then distributed to various different areas in the brain responsible for processing visual information. On completing the final signal exchange at the frontal lobe in the front of the brain, you will now "cognize" the existence of an apple in front of you. The details of this cognitive process are unknown. It is assumed, however, that you "cognize the apple" when a perceptual representation of the apple matches the memory representation of an apple in your brain. It is also assumed that when the memory representation of an apple is activated, you visualize or imagine an apple. It is assumed that the representation of an apple exists in the human brain in some form or another. Where does this representation exist?

Functionalists admit the existence of representation, whereas behaviorists do not. Behaviorism excludes subjective matters such as consciousness and the mind and therefore does not discuss representation as a means of explaining consciousness and the mind. It also denies the idea of a "central control tower" of any kind that controls consciousness and the mind to work properly as an intermediary of representation.

Denial of the central control tower originally comes from negation of the "homunculus." It was once believed that a very small human being called a homunculus lived inside the brain of a human and controlled the person's consciousness and mind, and that the spiritual activities of humans were governed by commands given by the homunculus. If one agrees with the existence of the homunculus, it is then necessary to discuss the working of the brain of the homunculus, making for a circuitous discussion that goes around endlessly without reaching a conclusion. How to stop this regression is one of the serious problems related to consciousness.

Descartes explained that the pineal body at the center of the brain played the role of a central control tower, but this idea is denied by modern brain physiology. For these reasons, behaviorists seem not to agree to the idea of a central control tower.

Personally, I believe that although the idea of a central control tower can reasonably be denied, we should admit the existence of asynchronous parallel and distributed systems. According to physiological knowledge about the human brain, it is certain that information is processed by nerve networks and the relevant processing is done by asynchronous parallel and distributed systems.

Connectionism is a theory that attempts to create such asynchronous parallel and distributed systems using artificial nerve networks called neural networks. At present, however, because highly functional computers are available, we do not use asynchronous parallel and distributed systems but simulate similar functions using ordinary synchronous computers. The term "simulate" here means mimicking.

4.8 Mechanical Systems Without Representation

The mechanical system without representation is generally known as a speed governor.

Figure 4.9. Heron's steam engine.

The role of this machine is to maintain the constant rotating speed of a flywheel driven by a steam engine. A steam engine is a mechanical system that makes use of the force of steam input into a cylinder to push down a plug-shaped component called a piston. The rod-like component attached to the piston pushes down the flywheel to convert the vertical motion of the piston into rotary motion.

Heron, ancient Greek mathematician and engineer, first invented a steam engine in the form of a rotary device (Fig. 4.9). His steam engine could not keep the rotating speed constant.

James Watt (1736–1819), a Scottish mechanical engineer, invented this speed governor in 1774. To keep the rotating speed constant, the steam pressure should be basically constant. Actually, however, the rotating force is used to do some work, and therefore when the workload varies, the rotating force also varies even when the steam pressure remains constant. For example, when a steam locomotive running across flatlands starts climbing a slope, it will slow down if nothing is done. This is called load fluctuation in physics. What should be done to maintain a constant rotating speed? The answer is to control the steam pressure relative to the load fluctuation. Watt's speed governor exactly satisfied such requirements for pressure control.

In 1774, steam engines were a practical mechanical system for delivering power safely and stably and contributed to the Industrial Revolution. Since then, a large volume of materials have been easily transported using steam locomotives and steamships to support the explosive development of industry.

Let us go back to the story of representation.

A researcher says that representation does not exist in Watt's speed governor. Is it true that his speed governor does not have anything that corresponds to "images of external objects that emerge on the consciousness" as representations? One point needs investigation. The metallic balls, the central components of Watt's speed governor (Fig. 4.10), spin outward from the central rotating shaft by centrifugal force as the rotating speed of the steam engine increases. At this time, the angle between the metallic arms holding the balls increases. When the angle of the balls increases, the steam valve is gradually closed to decrease the rotating speed. Conversely, when the angle decreases, the steam valve is opened to increase the rotating speed. The angle between the ball arms represents the internal condition of the steam engine, and we may call it a representation.

Figure 4.10. Watt's speed governor.

The researcher referred to in the previous paragraph denies this thinking. He denies it because, he says, what is considered a representation is not actually used in the steam engine. He argues that the angle is not used as a numerical value.

The author admits that the researcher is logically right. Indeed, the story of the speed governor seems to suggest the existence of "systems without representation," as they are called in behaviorism.

Since what determines the performance of this autonomous system is the amount of return and the time lag, it suffices in essence if the autonomous system works as expected irrespective of whether the existence of representation is agreed with or not.

In fact, we have today digital systems that use representations that function in the same way as the speed governor.

I think this discussion on the speed governor will eventually lead to the question of whether analog or digital control should be used.

Is this the reason why this is an essential discussion?

Speed governors incorporating a computer that is an aggregation of representations are already being used in real vehicles to control their engines.

The author therefore believes that, in essence, representations may well be used if and when they are required in the study of consciousness and the mind.

4.9 Qualia and Mirror Neurons

Qualia has been talked about much lately. In short, it refers to the feel of a material. David J. Chalmers said that qualia were the hardest problems in consciousness (Chalmers, 1996). He said so because behaviorists agree on the existence of qualia, but it is impossible for them to create, in a behavioristic manner, the qualia that appear in consciousness. It is far easier for functionalists to explain, as they need only to declare that a function to generate qualia exists in the brain. Of course, there still remains a problem in that even functionalism is unable to answer the question of how qualia are created.

The discovery of mirror neurons is described here. Mirror neurons have a special function and were discovered in 1996 by

(a)   (b)

Figure 4.11. Mirror neuron experiments. Cited from the research of Vittorio Gallese (1996).

G. Rizzolatti, an Italian brain science researcher (Rizzolatti, 1996). Mirror neurons were accidentally discovered when a researcher was having his lunch during an experiment and happened to notice that the nerve cells of the monkey in the experiment were activated vigorously. These particular nerve cells were also activated when the monkey was eating its own food. Surprisingly, the brain cells in question reacted in the same way as if the monkey itself was eating by just seeing others eating (Fig. 4.11). Based on this finding, the brain cell was named mirror neuron because it acts like a mirror.

Kenichiro Mogi describes how he was shocked on hearing the news of the discovery of mirror neurons. "Mirror neurons were a leap over forward-thinking and system theory-oriented researchers, and surprised many people by achieving the highest-level unification of connecting motor information on the 'behavior of the self' to sensory information on the 'behavior of the other.'" He continues, "Mirror neurons connect the actions of both the self and the other as if they were reflected in a mirror. As a result, for example, 'The other is doing this action, so if I were doing the same action, I would feel like such and such, and therefore, the other must also be feeling the same way.' The discovery seems to suggest that mirror neurons work as a nerve module that supports the ability to suppose the mental condition of others (the theory of the mind)" (Mogi, 2003). Mogi assumes that the mirror neurons in monkeys are closely associated with their behavior of pretending to be like humans.

The next topic is affordance theory and autopoiesis theory, which are currently attracting general attention.

4.10 Affordance and Autopoiesis Theory

Affordance is explained in detail by Toru Nishigaki (Nishigaki, 1999).

American experimental psychologist James Gibson introduced the theory of affordance in the 1960s. Affordance is an attempt to understand the problem of human consciousness and the mind from the viewpoint of interactions between the human and the environment. According to affordance theory, an object in the environment acts on ("affords") the human (the subject), thereby allowing the human to understand the object. For example, when an individual looks at an object that is a baseball, the individual is given the meaning from the baseball that "this is something for playing with that is thrown, caught, and hit." When a dog sees the baseball, the dog is given the meaning from the baseball that "this is something for playing with and to be chased after and bitten."

The affordance theory asserts that the meaning of an object in an environment is defined by the interaction between the subject and the environment. According to affordance theory, the meaning of an object is not subjective information arbitrarily interpreted by the subject based on stimuli received from the environment, but the meaning is instead information existing in the environment itself, which is objective information described by the environment and is simply transmitted to the subject. On the basis of this thinking, the affordance theory intends to find a solution to the problem of subject and object that is as objective a solution as possible.

The autopoiesis theory closely resembles the affordance theory. It was introduced by Francisco Varela (1946–2001).

The difference between affordance and autopoiesis is that with the former the meaning of an object is the objective information given by the environment, whereas with the latter that objective information constructs, or invents, the meaning of the object using experiential knowledge already learned within the subject.

4.11 Embodied Cognitive Science and Symbol Grounding Problem

This section introduces embodied cognitive science, which has been attracting the attention of many robot researchers today.

Embodied cognitive science is a belief that intelligence can only be understood in combination with physical systems. This concept captured the hearts of many researchers because it specifically proclaims that intelligence is in need of an embodied mechanical system. Embodied cognitive science represented a large flow of research covering the major achievements of robot researchers in the 1980s, which was later summarized by Rolf Pfeifer, Federal Institute of Technology Zurich, Switzerland, and other researchers (Pfeifer, 2001).

Pfeifer and other researchers considered that the study of artificial intelligence and robots had been critically blocked by two major obstacles, and he tried to achieve a breakthrough using a new keyword: embodiment. These obstacles are the frame problem and symbol grounding problem.

"Frame" refers to a symbolically represented knowledge database, and the term was originally proposed by Marvin Minsky. He declared that human intelligence could be realized on a computer using symbolic inference within two or three years. Certainly, artificial intelligence using symbolic inference contributed to the development of the **MYCIN** medical **expert system** designed to examine and diagnose infectious diseases and a robot simulator called SHRDLU (previously mentioned), which performed intellectual tasks as instructed by humans although in the world of building blocks.

In 1997, the IBM research group developed Deep Blue, a chess-playing computer. Readers may remember that the machine played chess with world champion Garry Kasparov and won a five-game match by two wins, one loss, and two draws. Deep Blue symbolized and processed the chessboard and the positions of the chess pieces and anticipated the next moves of its opponent by symbol-based reasoning.

There is no doubt that symbolic inference is one of the basic techniques for achieving artificial intelligence, but the frame problem prevented the results for use in the future possibilities of artificial intelligence.

John McCarthy is said to be the first person to point out the frame problem (1969). The essence of the frame problem is this: It is an illusion to believe that the knowledge database describes everything, as it is obvious that knowledge databases cannot describe the entire mass of knowledge. A nuclear reactor

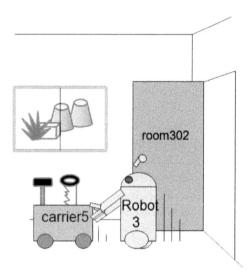

Figure 4.12. Frame problem.

maintenance robot, for example, operates on a knowledge database
that describes the structure and functions of the nuclear reactor.
Considering an extreme case, let us assume the reactor were
damaged by an earthquake or other accident; then the robot would
not have been able to use the knowledge database and therefore
would not move and do anything (Fig. 4.12). Generally, there
is another problem in that working environments are changing
continually and this is not a simple problem that can be solved by
just modifying parts of the knowledge database and updating it for
the robot to cope with changes in its environment. This is the frame
problem: the robot is required to respond to the situation using
"common sense," but this is unknown to robots.

The other problem is the symbol grounding problem. It is said
that Stevan Harnad first mentioned the symbol grounding problem.
The essence of the symbol grounding problem is how to define the
relationships between symbols used for processing intelligence and
real things in the environment (or "mapping"). Mapping does not
constitute a problem when humans stand in between the "symbols
in the robot" and the "environmental stimuli to be mapped by such
symbols," and interpret the relationships between them. Mapping

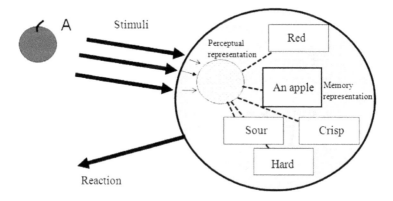

Figure 4.13. Symbol grounding problem.

between the environmental stimuli and the symbols in the robot is very difficult when the robot processes symbols while moving autonomously and interacting with its environment. For example, it is not yet defined how the environmental stimuli of object A in the real world, on the one hand, and Apple, Red, Sour, and other symbols (as memory representations) stored in the robot, on the other, should be automatically mapped onto one another. This is called the symbol grounding problem (Fig. 4.13). How does embodied cognitive science try to solve this problem?

The embodied cognitive scientists recommend not using, or minimizing the use of, knowledge databases in order not to be bothered with this problem. Alternatively, they recommend giving the feature of "situatedness" to the robot so that the robot can interact with its environment by itself without human or other intervention. Situatedness means that the robot is capable of acquiring information on the current conditions of its environment (or being situated) through interactions with the environment using its sensory sensors based on the embodiment. Considering these points, embodied cognitive scientists use "complete agents" as a tool for studying embodied cognitive science. A complete agent is a virtual or real robot featuring adaptivity, autonomy, self-sufficiency, embodiment, and situatedness. Such a robot is required to stay alive in the real world by itself.

Embodied cognitive scientists call a robot with such complete agent features a "non-Cartesian" robot to differentiate it from conventional robots. This is because conventional robots lack the property of adaptivity, that is, being able to act by adapting to any unknown environment, which one of the features of a complete agent.

Cognitive scientists further argue that a complete agent featuring such properties is capable of "emergence." Emergence is the capability of a robot to adapt to unknown environments unforeseen by the agent designer or developer and behave unexpectedly. These robot behaviors are not included in the internal characteristics of the agent nor input into it as a program.

There is a story that illustrates this idea in an easily understandable way: a parable of an ant (Fig. 4.14).

An ant is walking on a beach, cleverly avoiding rocks and water. Its walking motion might be scientifically analyzed but would require unbelievably complex formulas to describe the optimum

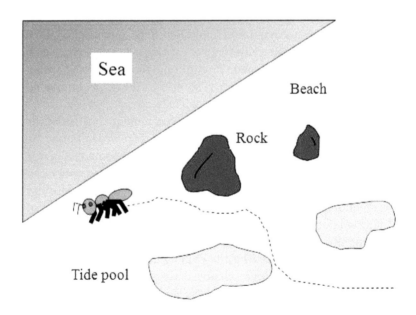

Figure 4.14. Behavior of an ant.

path for the ant to take. Researchers in embodied cognitive science doubt that this approach is basically misleading. If this approach were to be taken, we must first focus on the environment of the beach. It would be almost impossible, however, to scientifically formulate this kind of environment. Given this fact, the researchers realized that a new concept was required. They noticed that the walking motion of the ant should be formulated as an interaction with the environment at all times. Thus, the formulation would read in essence, "the ant takes a certain basic action under certain conditions." In this series of processes, "detecting conditions" and "taking basic action" would be repeated continually. In a sense, this is a very beneficial concept in designing an agent.

It is obvious that researchers in embodied cognitive science consider consciousness and the mind as a problem of emergence.

When, for example, a robot (an agent) is designed to avoid red blocks but approach green ones, it is possible to think that the robot hates red but likes green.

Certainly, from a behavioristic perspective, consciousness and the mind of other people "seem to exist," and in this sense behaviorism is the same as the complete agent.

However, I am clearly aware of my own consciousness (I possess self-consciousness), and I would like to emphasize that embodied cognitive scientists have not yet provided a clear explanation for this phenomenon of self-consciousness. Basically, however, I think their techniques — creating actual moving robots, realizing complex actions using a chain of basic actions, considering an action an interaction with the environment, and devising emergence-oriented designs — should be highly valued. I think, however, it is still difficult for them to elucidate the problem of "self-consciousness as being conscious that the self is conscious." I believe that we should further proceed with the study of human consciousness and the mind using the phenomenological approaches introduced by Merleau-Ponty.

Of course, they as "non-Cartesian" would accuse me of being a Cartesianist, but I would like to further study the problem of consciousness and the mind by interpreting science more broadly.

Let us now talk about consciousness and about the mind.

4.12 More Knowledge about Consciousness and the Mind

Tadashi Kitamura enumerates 10 properties of consciousness based on the Husserlian phenomenology (Kitamura, 2000). Following are the ten properties:

1. First-person property
2. Orientation
3. Relationships between action and result; the duality of consciousness
4. Expectation
5. Functions of determination and conviction
6. Embodiment
7. Consciousness of others
8. Emotional thought
9. Chaos
10. Emotion

These properties will be described in detail later in this book.

Other important thinking on consciousness is introduced now.

Naoyuki Osaka, Kyoto University, holds that consciousness has a three-layer structure (Osaka, 1996). His theory, which has already become established, is described as follows:

The lowest level is **Awakening**, the intermediate level **Awareness** and the highest level **Recursive Consciousness**. Awakening indicates that one is awake. A living creature apparently loses consciousness when sleeping and engages in biological activities when it is awake. Awakening refers to the state of being awake and performing biological activities. Awareness refers to the state of being awake and with consciousness directed at a certain object. The state of consciousness being intentionally directed at a certain object is called "orientation of consciousness." An example of this is a student who listens to a teacher speaking while taking notes in a notebook.

I would like the reader to learn about **reversible and ambiguous figures**, which are famous in Gestalt psychology to familiarize yourself with the nature of consciousness.

Figure 4.15. Rubin's cup.

When viewing these examples, your consciousness is at the intermediate level, and you will clearly understand what is meant by "orientation of consciousness."

The Rubin's cup is a famous example of a reversible figure (Fig. 4.15). Two meaningful images exist in the figure and appear alternately: two human faces facing each other and a bottle.

The Kanizsa triangle is a typical example of an ambiguous figure, an optical illusion.

Gestalt psychology is a theory that attempts to understand the human mind as a holistic structure, or *Gestalt* in German, rather than as a collection of individual functioning units.

Recursive consciousness refers to the state where consciousness performs activities directed toward consciousness of the self. This is called self-consciousness.

Humans, thus, can be conscious of what they are currently conscious of. Humans are inherently bestowed with a kind of metaconsciousness or superconscious.

Osaka says that metaconsciousness is the function that makes humans truly human-like.

From the perspective of functionalism, a supersystem that overviews a system is foreseen to exist at a higher level of the system.

During his experiments, brain scientist Wilder Penfield directly touched the cerebral cortexes of patients with an electrode during surgery to learn what phenomena would occur when such stimuli were given. The experiments were performed for the purpose of minimizing loss of brain functions due to unintended excision of part of the brain in brain tumor resections. It is said that the patient does not feel pain because there are no sensory cells in the brain.

Penfield asked the patient in the experiment, "Begin counting out loud from the number 1." When he stimulated a certain part of the brain while the patient was counting, suddenly the patient could not say the numbers any more. Another patient was said to have told the doctor that he vividly remembered events from his childhood days when a certain cortex was stimulated.

Now we will revert to the topic of self-consciousness. When a patient told Penfield, "I am now being treated by you," Penfield tried to identify the location but he failed to do so.

As mentioned before, Descartes declared that only humans possessed consciousness. This would mean that dogs and cats have no consciousness functions.

Of the three levels of consciousness advocated by Osaka, I think awareness, the intermediate level, may exist in dogs and cats, but it is doubtful if these animals have self-consciousness. Cats chase rats by orientation, and dogs chase and catch a ball tossed to it by humans. These facts clearly show the existence of awareness in these animals. Dogs and cats, however, do not seem to respond to their mirror image very well. There is no evidence to show that they recognize their images reflected in a mirror. It is, therefore, assumed that self-awareness is not a function that exists in dogs and cats.

An inspection method is available for determining the existence of a high-level cognitive function. It is a method to check if the self - image reflected in a mirror is cognized or not, and it is known as the mirror test as proposed by Gallup.

Will anthropoid apes that resemble humans pass the mirror test?

There is evidence to believe that chimpanzees possess self-awareness. A chimpanzee was given a mirror and an opportunity to play with it for a sufficient length of time. Later, the chimpanzee was surreptitiously marked with signs on its forehead with non-irritant dyes. When once again given a mirror, the chimpanzee touched the marks on its forehead while looking into the mirror. This indicates that the chimpanzee cognized that the image reflected in the mirror was an image of itself. The function to cognize one's own mirror image is self-awareness; so we can determine that chimpanzees possess self-awareness.

Besides chimpanzees, there have been reports that dolphins, Indian elephants, and magpies also pass the mirror test.

There is an interesting story about consciousness. It is an observation about **free will**. What makes humans truly human is the free will that humans possess. Everyone believes that he or she can freely decide matters in any place and at any time. An unbelievable fact was revealed in experiments.

The human subject in the experiment was instructed to bend his fingers at any time he wished. The observer was tracing the activity of the subject's brain with an electroencephalograph (EEG). The subject bent his fingers at a certain time T. The brain started its activities, however, about 1 second earlier than time T. This is interpreted as follows: Although the subject thought he made the decision to bend his fingers and did so at time T, a preliminary decision had already been made about 1 second earlier in the brain. This famous experiment was conducted by H.H. Kornhuber (1976) in Germany. Later, researchers, including Benjamin Libet, Professor Emeritus at the University of California, San Francisco (UCSF), performed additional detailed experiments to verify Kornhuber's findings in the 1980s.

These experiments seem to lead to the conclusion that humans do not have free will because the brain started to act before the human noticed that he actually bent his fingers, and that the human is never able to notice this preparatory work of the brain. Is it true that humans really do not have free will?

This problem will be discussed in detail later in this book.

4.13 Summary and Observations

I would like to point out, in the first place, that the study of consciousness and the mind has a long history extending from olden times to the present day.

Nevertheless, the existence of consciousness is still described as a subjective sensation and as such we cannot advance further.

I think it is generally accepted as an objective fact that consciousness exists in the left hemisphere of the human brain. This is supported by anatomical findings in that if the right hemisphere of the brain is surgically excised, the consciousness function of the patient is not critically affected, but if the left brain is removed the consciousness function may be totally lost depending on the excised areas.

There is no objective analysis of the effects of excision on the consciousness of the subject. Only the subjective judgment of the experimenter is relied on to evaluate the results of the experiments.

I do not agree, however, with the argument of behaviorists that humans possess no consciousness. This is because I can feel that I have consciousness. This feeling is, without a doubt, a subjective sensation. I should note that this subjective feeling is not necessarily irrelevant to physical processes.

In other words, if the feeling of being conscious passes through a physical process, we can scrutinize the process from the perspective of physical science.

It is also possible that the function of human consciousness is unique to each individual. If these individually unique functions of consciousness are based on physical processes, we can analyze them scientifically. By comparing these individually unique physical processes, we may find some typical and common physical features among them.

Since phenomenologists seem to have arrived at a certain common recognition about consciousness and the mind, we should not doubt that someday we will be able to find an objective physical process that describes the occurrence and action of consciousness and the mind.

What is certain at this time is that the existence of consciousness and the mind is known only to oneself. With regard to others, we only feel that they must also have their own consciousness and mind like me.

Descartes insisted on the consciousness of the existence of the self when he declared, "I think, therefore I am." For us today, however, it is totally unknown if Descartes possessed consciousness, and what kind of conscious activities led him to say that he did.

Nevertheless, it is true that his discourses significantly affect our conscious activities.

Chapter 5

Professor Valentino Braitenberg's Vehicles

Valentino Braitenberg was director of the Max Planck Institute for Biological Cybernetics, Germany, where, using small robots equipped with sensors and drive motors, he studied robots with a mind from the standpoint of functionalism.

In brief, functionalism is a method to try to understand something by analyzing the function inherent in the thing to the extent possible by human wisdom. What is important in this definition is that the understanding is limited to the range possible by human wisdom.

Braitenberg was one of the pioneering researchers of robotics, and he designed robots by transmitting information directly from sensors to drive motors. He invented many vehicles, each with successive improvements, and showed that a wide range of robots from toy-level to a self-conscious robot could be made using very simple components (Braitenberg, 1987). He was a forerunner in research on evolutionary robotics and embodied cognitive science, which are currently very actively studied in the field of robotics. Many of the methodologies adopted by these researchers had already been suggested by Braitenberg. This fact is indeed marvelous.

Creation of a Conscious Robot: Mirror Image Cognition and Self-Awareness
Junichi Takeno
Copyright © 2013 Pan Stanford Publishing Pte. Ltd.
ISBN 978-981-4364-49-2 (Hardcover), 978-981-4364-50-8 (eBook)
www.panstanford.com

For instance, the idea of a modern evolutionary robot is the same as what Braitenberg called a Darwinian robot. A robot that expresses the feelings of "like and hate" using the connections between the sensors of artificial neural networks and drive motors, a phenomenon that would later be called emergence, was also conceived by Braitenberg. I believe that his research should be further evaluated. This underestimation seems to be due to the argument that "he sides with functionalism and admits the existence of representation," because many scientists favor the behavioristic stance. His vehicle robots are introduced now.

5.1 Braitenberg's Vehicles 1 Through 4

Braitenberg's vehicle 1 has the simplest structure. It consists of a sensor and a drive motor. The sensor and the drive motor are connected by an artificial nerve. The motor drives a wheel located at the rear of the robot (Fig. 5.1a). The robot is just like a small cell swimming with a flagellum. Vehicle 1 does not have a rudder. When it runs, it always runs straight ahead. A robot is almost always asymmetrical, so that the robot may make a slight turn. The sensor can be any sensor. Let us assume a temperature sensor. The sensor emits zero signals at the absolute zero point. At other temperatures, the sensor issues stronger signals as the temperature rises.

The signal is transmitted to the drive motor via the artificial nerve. The motor's speed (rotation) increases or decreases with the strength of the signal. Put this robot in a space, and it detects the ambient temperature and starts running at a speed corresponding to the detected temperature. The robot slows down in areas with lower temperatures.

To a casual onlooker, this robot looks like a simple life form such as an amoeba that becomes more active in places of high temperature.

When the reaction property of the artificial nerve is designed as shown in Figure 5.1c, the artificial life will be activated at a certain temperature.

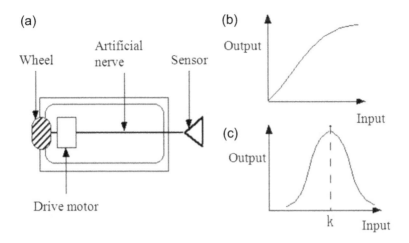

Figure 5.1. Braitenberg's Vehicle 1.

Braitenberg's vehicle 2 has two sensors that are connected to two respective motors. There are two types of artificial nerves: parallel (Fig. 5.2(1)) and crossed (Fig. 5.2(3)). The plus and minus signs (+ and −) indicate the property of the artificial nerve. For a plus (+) sign, the output of the nerve increases with increasing output signals of the sensor (facilitation). For a minus (−) sign, the output of the nerve decreases with increasing output signals (inhibition). Only the minus sign is shown here, but there should be no confusion caused with the plus sign omitted.

Let us assume that this robot is equipped with light sensors. The robot in the figure, with a light just in front of it, will go straight toward the light with its speed increasing until it crashes into the bulb and destroys it (Fig. 5.2(2)a).

If the light is not just in front of the robot, the sensor closer to the light is activated more than the other sensor, and the relevant motor increases its speed. As a result, the robot follows a path to avoid the light (Fig. 5.2(2)b). If the robot sees the light just in front of it again, the robot will rush toward the light.

The robot shown in Fig. 5.2(2) is a fearful and aggressive type since it avoids light but suddenly attacks it depending on the position of the light.

Figure 5.2. Braitenberg's Vehicle 2.

The robot shown in Fig. 5.2(3) is identical to the robot shown in Fig. 5.2(1) except that the nerves are crossed. This robot also goes straight ahead and collides with the light if the light is just in front of it. If the light is on the left side of the robot, the drive motor on the right side increases its speed, an opposite action because the nerves are crossed, and the robot turns toward the light.

Like a guided missile, the robot rushes at the light as if it were being sucked into it (Figure 5.2(4)). This is an aggressive-type robot.

The activation of the nerves is inhibited in vehicle 3.

In Fig. 5.3(1), the right and left nerves are in parallel, and their activation is inhibited (note the minus sign). This robot, placed at an angle to the light, slows down as it draws closer to the light and finally stops because the sensor on the side facing the light emits a higher signal than the other sensor and speed of the motor on the side facing the light slows down (remember the nerves are parallel and inhibited). The motor on the side away from the light runs faster than the motor on the side facing the light, and thus the robot turns itself toward the light (Fig. 5.3(2)). Both motors slow down as the

Figure 5.3.   Braitenberg's Vehicle 3.

robot approaches the light, until the robot stops before reaching the light (Fig. 5.3(2)a).

The robot shown in Fig. 5.3(3) has crossed nerves and the signals are inhibited. If the light is just in front of the robot, the robot goes straight toward the light while its speed continues decreasing, and it stops before reaching the light (Fig. 5.3(4)b). Otherwise, the robot turns away from the light while slowing down and gains speed as it moves away from the light (Fig. 5.3(4)c).

The robots that slow down and stop near the light may be said to be "affectionate" toward the light, whereas the robot shown in Fig. 5.3(3) is "fickle" because in most cases, it visits the light here and there but quickly runs away.

Braitenberg's vehicle 4 moves around the light in various patterns, reflecting the reaction properties of the artificial nerves Braitenberg devised.

Vehicle 4 seems to be devised with its own value and liking when we see how it approaches, stays, and leaves the light depending on the built-in reaction properties.

Braitenberg assembled logic functions in the nerves of vehicle 5.

By combining the artificial nerves, logical sum (or), logical multiplication (and), and negation (not) can be achieved in a robot. This will be discussed in detail in Chapter 7.

An adder is constructed when these three logical computations are available. Once addition is possible, subtraction, multiplication, and division with the remainder can be calculated. For example, multiplication is done by repeating addition, subtraction by adding complements, and division with remainder by repeating subtraction.

In short, a computing machine can be constructed by combining artificial nerves. This means that vehicle 5 can compute anything that can be calculated on a computing machine. Memorization is also possible. For example, the robot will count and remember the number of right turns it made during a journey.

Next, Braitenberg attempted to create a robot with a brain.

5.2 Robot with a Brain

Braitenberg proposed what he called "the law of uphill analysis and downhill invention" (Braitenberg, 1987). This means that analysis is difficult like climbing a hill, but invention is easy like going down a hill. Braitenberg does not mean invention is easy, but says that invention is easier than analysis.

Specifically, it is easy to say something about the behavior of a robot by looking at it, but it is difficult to explore the internal operation of the robot that caused it to perform a certain behavior. It is very much analogous to the complexity of analyzing an ant's behavior cited before. This also applies to humans. We can easily see what a person has done, but it is difficult to analyze the person's principle of behavior.

Braitenberg did not mean to give up exploring the mystery of human behavior. No scientist would say so. On the contrary, he urges us to build robots to assist us in researching the facts underlying human behavior. He suggests that we try to create the functions of human brains through an engineering approach and elucidate the mystery of the brain using the knowledge derived from such efforts. He studied, for example, the problem of will by analyzing the behavior of robots.

Consider vehicle 2, in which the amount of light shed on the sensors is very small. A robot, being a physical mechanical system, always has problems peculiar to physics such as friction and force of inertia. It is possible that even if the sensor detects light, the wheels might not turn or the robot might remain standing still due to friction. But at any moment, the robot might start moving with slight change in the environment or slight variations of unstable elements of the mechanical system related to friction. A researcher who happens to see the robot move suddenly may think that the robot has made up its mind to move. This is because the robot, which has been standing still, suddenly starts to move by itself. The researcher may conclude that a complex process of will exists behind this robot's decision making. Actually, however, there was only a slight change in light intensity.

The decision-making process occurring in the human brain may not be as simple as that. Nevertheless, it is probable that just a tiny amount of something affects the decision-making process going on in the brain. No one can deny that these trials would be a clue to reveal the truth.

At this point, philosophers may cite *Critique of Pure Reason*, written by Immanuel Kant (1724–1804) (Fig. 5.4). Kant says in this writing, "Indeed we may have found clues. But they are based upon the human recognition and therefore cannot be anything beyond human recognition." This interprets into, "Humans can recognize only that which they can recognize." This may seem a matter of common sense, but actually a profound meaning is hidden.

In *Critique of Pure Reason*, Kant mentions that the existence of God, the existence of the Soul, and the existence of free will cannot be known by humans. To advocate his agnosticism, Kant argues that "it is possible to prove the existence of God and also possible to prove the nonexistence of God" and holds that there exist matters that humans cannot know.

Immanuel Kant
1724–1804

Figure 5.4. Immanuel Kant.

Agnosticism is the belief that reality behind sensory experiences given to consciousness cannot be recognized demonstratively (*Kohjien*).

The author understands agnosticism but believes that excess reaction is not necessary. It is too extreme to believe that we cannot know anything just because we recognize agnosticism.

Consider clocks, for example.

According to agnosticism, the conclusion is that you cannot build a clock that shows the correct time.

In ancient Greece, Heron improved on water clocks. Galileo Galilei (1564–1642) invented a pendulum clock using the isochronism of a pendulum. The balance mechanism with gears started to tick regularly. Nautical clocks were then invented, which were capable of compensating for errors caused by the rotation of the Earth. This was made possible by the invention of a mechanism in which the balance itself turns. Modern clocks include quartz clocks that work on the oscillation of a quartz crystal, and then there are atomic clocks. Despite all of these developmental achievements, correct clocks are still to be devised. Nevertheless, we have never experienced any practical problems with the balance mechanism of mechanical clocks for hundreds of years. Mechanical clocks are once again popular today (Fig. 5.5).

Figure 5.5. Revival of mechanical clocks.

This concept is called pragmatism.

Pragmatism is "a way of thinking about things in concrete based on events (*pragma* in Greek); the meaning and truth of a concept is evaluated in terms of how it works and its consequences" (*Kohjien*). William James (1842–1910), an American, was an ardent advocator of pragmatism.

Some readers may claim that pragmatism tends to give rise to the "arrogant overconfidence of scientists" (Friedrich Wilhelm Nietzsche, 1844–1900) and includes the risk of bringing about the collapse of the world at any time.

It should be possible for humankind, I believe, by cautiously avoiding "arrogant overconfidence" and without overestimating the difficulty of "brain problems," to gradually increase its wisdom to such a level where there is no practical problem just as exemplified by the developmental history of clocks.

The author's comments on agnosticism and pragmatism have been introduced here to defend Braitenberg's belief.

5.3 Idea of an Evolutionary Robot

Braitenberg prepared a table, and selected and placed on it some relatively excellent robots from those that he had made. Various components used for the development of the robots were also placed on the table.

The robots moved around on the tabletop at random, with some falling off the table. With the fallen robots left as they were on the floor, new robots were duplicated on the table using the available components. When duplicating the robots, any errors that may have existed in the original robot were also incorporated in the new robot. By repeating this procedure, vehicle 6, one that would never fall off the table, should be created. This vehicle is called a Darwinian robot (Fig. 5.6). This was named after Charles Darwin (1809–1882), who introduced his theory of evolution.

The engineering technique to produce evolutionary robots that simulate the evolution of organisms is described in Chapter 8.

I do not insist that the evolutionary robot was first conceived by Braitenberg, but it should be noted that he explained that a brain can be built in a robot by actually combining the theory of evolution and a mechanical robot.

5.4 Vehicles 7 Through 10 with Associative Concept

Braitenberg's vehicle introduced a wire M that has a high initial resistance that decreases when both parts connected at either end of the wire are activated. The resistance returns to the initial value with time. Braitenberg calls these wires "mnemotrix."

Various sensory motor modules of the robot are connected together by the Mnemotrix wires. Hereafter, my description may differ slightly from Braitenberg's writing, but the details will be explained by following Braitenberg's ideas.

Figure 5.7 shows the vehicle 7 robot with logic artificial neural networks. The robot is provided with color sensors "a" and distance sensors "b." Drive motors are indicated as "c." The robot has two sensor drive modules, a-c and b-c. Wires M connect the artificial neuron units "d" located in the middle of the system. These intermediate units integrate the result of cognition of both the color and distance sensors and output the result to the drive motors.

Assume that this robot has learned to regard a rapidly approaching red object as dangerous. The resistance of the wires M

Repair and return

Figure 5.6.   Idea of an evolutionary robot.

connecting the intermediate unit of the a-d module representing "red" and the intermediate unit of the b-d module representing "rapidly approaching" is slightly lower than the other wires.

As a result, even when only the color information of "red" is given, the robot behaves evasively because the "red" information stimulates the b-d module representing "rapidly approaching" via the wires of slightly lower resistance. Braitenberg interpreted this as the "judgment of 'red' provoked the judgment of 'rapidly approaching' by association."

Concepts arising from association are called associative concepts. It is also possible, conversely, that the judgment of "rapidly approaching" invokes the judgment of "red" by association. By the

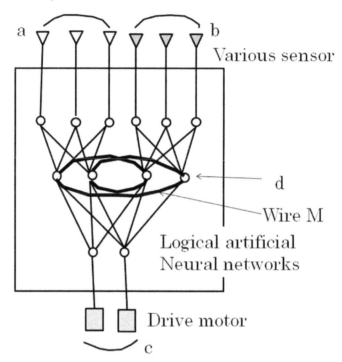

Figure 5.7. Vehicle 7 with associative concepts.

above statement, Braitenberg admits the distinct existence of the concept and the representation of "red" and "rapidly approaching" in the intermediate units, i.e., in the robot. He also explains that wires M created a new concept of a "red and rapidly approaching" object.

According to Braitenberg, if a robot that has learned an association between "green and rapid approaching" behaves evasively by chance for a "blue" moving object (a color that the robot has not yet learned), it is possible for the robot to abstract colors (make a distinction between colored and colorless robots) and the accompanying generalization.

Abstraction is a "mental action to extract and grasp certain aspects and properties of things or representations" (*Kohjien*).

Generalization is the "forming of general concepts or laws by shaking off specifics and saving what is common."

Braitenberg introduced his vehicles 8 and 9 to describe techniques to conceptualize a robot's environment, such as space, objects, motion, and form, within the robot using vision sensors.

The discussion of a robot's learning of its environment is omitted here because in Chapter 7 it is explained that artificial neural networks are capable of such learning and, further, this topic is irrelevant to the problem of the mind.

Braitenberg's vehicle 10 was designed to study the problems encountered when forming a new concept while keeping past associations. He proposes to constantly circulate stimuli for the associations that were memorized in the past.

5.5 Robot with a Sequential Concept

Braitenberg's vehicle 11 features new wires (E) that he calls "Ergotrix." These wires E are used to memorize the order of the occurrences of representations "a" and "b." For example, one may wish to have the robot remember to turn right when "an obstacle pops up in front of the robot immediately after the red lamp lights" (order of occurrence of two events). Wires E connect the intermediate nodes "d" to one another as shown in Fig. 5.8. For any two intermediate units n_1 and n_2 that are connected by wire E, signals are transmitted from n_1 to n_2 when n_1 fires and then n_2 is ignited just a short time later. Importantly, the connections using this wire are different from a simple artificial neural network. The directionality of signal flow is assigned to each of these wires.

The robot shown in Fig. 5.8(1) comprises two sensor drive modules, a-c and b-c. Like the previous robot, sensory sensors "a" are color sensors and sensors "b" distance sensors. The difference from the previous robot is how the intermediate units "d" are connected to one another. New pathways are added to allow recurrence to the intermediate units themselves.

An example of a circuit network with wires E is shown in Fig. 5.8(2). Here, d_1 is the intermediate unit for the color sensor and d_2 for the distance sensor. In the circuit network, a single input

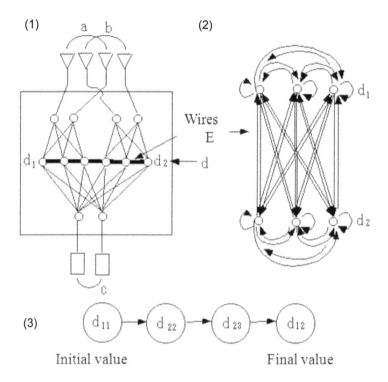

Figure 5.8. Vehicle 11 with a sequential concept.

circulates through representations d_1 and d_2, just like a series of variations of state occurring in an automaton. The robot is eventually expected to reach a final state as shown in Fig. 5.8(3). In Figure 5.8 (3), the robot memorizes the environmental changes as a temporal change of representations: d_{11}, d_{22}, d_{23}, and d_{12}. Specifically, the robot saw a red lamp, felt a wall on the right, then felt another wall in front, and, lastly, saw a blue lamp.

An automaton is defined as "a general term for mathematical models of various computing mechanisms including computers. A finite number of internal states are considered, connecting inputs and outputs. The internal state varies with inputs, or the output is a function of the internal states" (*Kohjien*).

This robot is capable of memorizing events occurring in the environment and the order of their occurrence. We then add wires M to wires E in Fig. 5.8. This robot shows us that a new concept can be formed by chaining representations in an orderly manner. A concept created by time-series events is called a sequential concept.

5.6 Vehicle 12

Vehicle 11 was improved, at one point, to become vehicle 12. A new type of wire C was added to control the threshold function of all artificial neurons that make up the logic unit of the robot.

Wires C run from the control box of the logic unit comprising the brain of the robot to all neurons in the brain to modify the threshold functions (Fig. 5.9).

"Modify" here means to alter the threshold values of all neurons according to the number of currently firing artificial neurons in the logic unit. If the number of currently firing artificial neurons

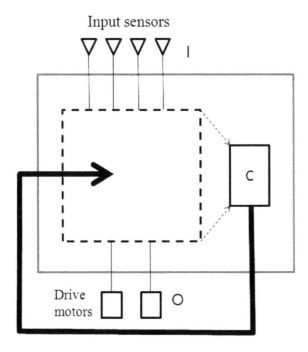

Figure 5.9. Threshold Control Circuit.

increased, all threshold values are increased, and as a result, the number of artificial neurons that will fire next time decreases.

The robot's brain accepts input pattern I at its logic unit and outputs pattern O from the logic unit. The output is under the influence of the threshold functions that are changed by the control box. The role of wires C is to highlight the currently represented concept against all others. By elevating the threshold values by controlling all threshold functions using wires C, it is possible to dispel all concepts but the strongest concept g in the logic unit. I personally believe that this function can be said to be a "concentration of attention in consciousness." As a result, concept g remains active with the strongest attentiveness of all concepts both spatially and temporally due to the working of wires M and E.

Concept g then generates a sequential concept g_1 using wire E, and this concept in turn creates several associative concepts g_{1i}. Note that at this time the firing of the artificial neurons increases sharply in the logic unit, and the threshold values are unanimously elevated due to the working of the control box. Thus, concept g_{15}, which is most active in the logic unit, is selected (assuming $i = 5$).

Next, a sequential concept g_2 of concept 15 arises.

We can safely expect that the concept arising at this time is a concept deeply embedded in the memory through past experience.

Braitenberg explains this continuous occurrence of concepts as "trains of thought." He asserts that this scheme is very similar to the human brain. With regard to the occurrence of concepts, Braitenberg says it is impossible to foretell what concept will occur next. This is because we cannot foresee how many firings might occur as a result of the threshold control at the logic unit according to function f. This means that you cannot predict the next occurring concept. The unpredictability of the number of firings is akin to chaos.

Chaos means "in a system where initial conditions determine the succeeding motions uniquely, a slight change in the initial conditions results in a significant difference after a long time, practically making it impossible to predict the end result" (*Kohjien*).

By observing the behavior of vehicle 12, Braitenberg concluded that this robot has free will. Behavior is not predictable not only for this robot but also for people. If we recognize free will to mean that no person's instruction can affect another's behavior and that this is one of the dignities of humans, then we must accept that the robot has free will.

5.7 Vehicle 13

As already discussed, Braitenberg's vehicle 12 generates endless concepts, and a train of thoughts appears on the robot. Despite these many concepts that occur, we do not see any voluntary orientation in the thinking of the robot.

Humans are said to think of a number of options in the course of attaining their final target. This is "oriented thought," which is the theme of this section. According to Braitenberg, there are two types of orientation: prediction and judgment of good and evil. To incorporate the prediction function, sequential concepts must be memorized via wires E. Sequential concepts are generated each time the robot behaves. In the prediction process, no synchronization with individual behaviors is required, so that concepts may be stored in the robot at high speed, and independent of the progress of behaviors. The device responsible for this process is called the predictor.

Assume the robot presently has a concept g_3 that is established. The predictor starts working from here, and generates a chain of sequential concepts $(g_3 \rhd g_{31} \rhd g_{32} \rhd \cdots \rhd g_{3m})$ using wires E. These concepts are what are expected to occur in the future, given the current conditions and past experience. It is unknown, however, whether the robot will really generate g_{31} after g_3 as predicted.

This is because the robot, moving in a variable environment, may develop a concept directly from the changing environment as detected by its sensors, that is, a concept irrelevant to the control given by wires M or E.

If the input received by the sensors is ignored, the robot will perform a behavior that generates the sequential concept g_{31} as predicted, and the sensors will be accordingly changed, also as predicted. This would be followed by a number of associative concepts $(g_4, g_5, \ldots g_s)$ via wires M, until the most strongly learned concept g_5, for example, occurs by virtue of the controlled threshold values. If a sequential concept occurs again, a new prediction process starts.

Braitenberg's strategy for cases in which the prediction fails to match the changing environment (as detected by the sensors) is to "believe in the sensors when in doubt," and he turns off the switch of the predictor.

Regarding the judgment of good and evil, humans need not teach robots, rather, according to Braitenberg, the robot will learn by itself that an object is good when the robot approaches it and is evil if the robot moves away from the object. This is indeed a simple solution but you may get it when you think that it is something like the intuition that is deeply embedded in the genes of humans.

5.8 Vehicle 14

Braitenberg's vehicle 14 is different from vehicle 13 in that multiple chains of sequential concepts $(l_1, l_2, \cdots, l_q)$ predicted from the current concept are proposed.

For example,

$$l_1 = (g_3 \triangleright g_{31} \triangleright g_{32} \triangleright \cdots \triangleright g_{3r})$$
$$l_2 = (g_3 \triangleright g_{41} \triangleright g_{42} \triangleright \cdots \triangleright g_{4r})$$
$$\bullet$$
$$\bullet$$
$$l_q = (g_3 \triangleright g_{k1} \triangleright g_{k2} \triangleright \cdots \triangleright g_{ku})$$

The robot selects the sequential concept that gives the best condition for itself. The evaluation criteria would be "advancing toward a better state for the robot." Looking at this robot from outside, we feel that it behaves as it likes with unwavering determination.

5.9 Summary and Observations

Braitenberg proposed many vehicles, from the simplest vehicle 1, through to vehicle 14, the most complex. All of his proposed robots basically consist of a logic unit comprising artificial neural networks that connect sensors and drive motors.

Braitenberg says that representations or concepts of, for example, "red" and "nearby" can be created in the logic unit by adequately stimulating sensors. He asserted that such formed concepts can be abstracted and generalized in the same logic unit.

Even a very simple robot looks like it is behaving "lovingly," "with hatred," or "aggressively" when we look at it.

Braitenberg's idea resembles the concept of an evolutionary robot. In other words, he believed that a robot capable of adapting to its environment could be created as though going through the selection of living organisms. He showed a method to generate associative concepts in a robot by incorporating wires M in the logic unit. He also showed a method to generate sequential concepts that occur as the robot makes a succession of behaviors by incorporating wires E in the logic unit. He then showed us a robot that achieved trains of thought by using wires M and E simultaneously. He then assembled a predictor that automatically generates sequential concepts without necessitating the occurrence of behavior in the robot using wires E and showed that the robot behaves by predicting the future.

Lastly, Braitenberg improved on the predictor and asserted that using the predictor, a robot would be able to "voluntarily move toward a better state for itself in the future" and that by selecting from multiple candidates that are expected to occur from a given concept, the robot could select the optimum chain of sequential concepts that would guide it toward a better state in the best way.

Braitenberg may be said to have created small robots with a mind from the functionalistic viewpoint of a robot researcher. He asserted his law of "uphill analysis and downhill invention." With those words, he confessed that knowing whether a mind really resides in a machine was as difficult as going up a hill, and he tried to find a clue to "uphill analysis" by inventing machines with functions similar to a mind.

Chapter 6

Professor Rodney Brooks' Robots

In 1986, Rodney Brooks, then a roboticist at the MIT Artificial Intelligence Laboratory in the United States, announced a new robot. This robot featured a novel design compared with conventional robots. Brooks' design methodology is known as subsumption architecture. Conventional robots are generally called mechatronics models, which Brooks called a Cartesian model. Figure 6.1 shows an example of the mechatronics model.

A robot captures information about its surrounding environment by "sensing." The robot then recognizes the environment based on the information it captured. For example, the robot recognizes that an obstacle is in front of it. On the basis of this recognition of the environment, the robot generates an action plan. On discerning an obstacle, the robot itself creates a plan to avoid the obstacle. The collision avoidance plan can include various actions. For example, the robot may stop, slow down to let the obstacle go by, increase speed to cut across in front of the obstacle, or somehow run away. It selects the action to take depending on the given situation. It then takes the selected action using the drive system. As a result, the robot moves as planned, but here a serious problem arises. The environment changes as soon as the robot moves. The robot discerns that its environment has changed via its sensors, and the process is

Creation of a Conscious Robot: Mirror Image Cognition and Self-Awareness
Junichi Takeno
Copyright © 2013 Pan Stanford Publishing Pte. Ltd.
ISBN 978-981-4364-49-2 (Hardcover), 978-981-4364-50-8 (eBook)
www.panstanford.com

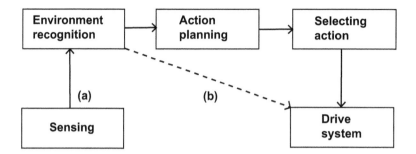

Figure 6.1. Mechatronics model.

repeated again (Fig. 6.1). A Cartesian robot thus repeats the process from sensing to behaving. There is no mistake in this process. All conventional robots are designed after this model.

However, many researchers, including the author, are not necessarily completely satisfied with the mechatronics model. Typically, no problems are experienced when researchers move a robot in a laboratory for research purposes, but this serial processing system turns into quite an annoying process once they attempt to develop a practical or commercial robot.

In the laboratory, researchers simply pursue the study to achieve their purpose and are not much concerned about the real-time response. In addition, we should take into consideration that the speed of computers was very much slower in the 1980s when Brooks conducted his robot studies. At that time, if one adopted an imaging system for the sensors, the turnaround time was terrible — as long as several seconds to complete the processing for a single screen. Nevertheless, researchers tried hard to improve the "real-timeness" by, for example, introducing natural constraints, devising bypass routing from environment recognition to the drive system, and incorporating hardware to execute certain programs (Fig. 6.1b).

Natural constraints include the well-known epipolar constraint explained in Chapter 2. In natural environments, for example, you cannot see an object behind an obstacle.

Brooks introduced his new robot design methodology against this historical backdrop. With the robot system conceived by Brooks, parallel processing replaced the serial processing of the

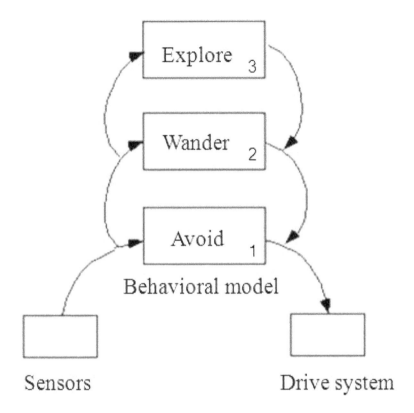

Figure 6.2. Subsumption architecture.

conventional mechatronics models. The structure of his system is shown in Fig. 6.2.

Apparently, one cannot see anything suggesting environment recognition or action planning. In fact, these two factors are actually hidden dexterously. Note that the behavioral modules are stacked one above the other. Each layer corresponds to an independent behavior program. Brooks' robot switches its behavior using the rule, "when a layer is unable to function, a lower layer will take over." For example, if exploring is impossible, the robot goes to and repeats wandering and avoiding (Fig. 6.2). Any layer functions with support from the underlying layers. "Subsume" means that a higher layer includes all the underlying layers. This is the meaning of "subsumption." The capacity of the behavior layers increases as

Figure 6.3. Brook's robot Cog.

one climbs the stack of layers. For example, the lowest layer is "avoiding," the one above it "wandering," and the highest "exploring." "Exploring" is selected only when there is no necessity of avoiding and wandering. To put it the other way around, avoiding is selected when exploring and wandering are suppressed.

The switching of the behavioral modules is triggered by replacing (or suppressing) inputs transmitted from the sensors to the behavioral modules and inhibiting outputs from the behavioral modules. The subsumption architecture conceals the process of environment recognition by using a dexterous technique of replacing inputs transmitted to the behavioral modules and suppressing outputs from them. It also hides the process of action planning by organizing a variety of simple behavioral modules into layers from the low to high levels. The net result is that the robot engages in continuous behaviors and will not end up in a deadlock thanks to the behavior-based design. The sensors and drive system are directly connected via the behavioral modules, so that the reaction time from sensing to behaving is short. Furthermore, the robot will never stall because multiple behavioral modules are prepared in layers and at least one or more modules are always available for functioning. It is easy to retrofit behavioral modules on already designed robots.

The robot design method proposed by Brooks has widely influenced later robot researchers.

Brooks has recently created a humanoid robot called Cog and has tried to attain in it a high-level intelligence similar to that of humans through learning by interacting with the environment and by adding behavioral modules. According to a recent progress report, it is getting difficult to increase the behavioral modules as the robot rises higher in the layers.

Summary

Prof. Rodney Brooks at MIT proposed a new robot design methodology called subsumption architecture. This design methodology may be said to be a parallel processing type as opposed to the conventional serial processing type of robot. The robot's reaction is quick because the sensors and drive systems are directly coupled via behavioral modules. Many behavioral modules are available and are stacked in layers one above the other from low to high levels.

Brooks believes that by adding relatively simple behavioral modules, high-level knowledge similar to that of humans can be realized. His approach is behavior based but is also functional because the robot's behaviors are classified by several functions. Brooks does not talk much about human mental activities but seems to think that these will emerge on their own as behavioral modules are added. If he thinks emergent human mental activities can be realized by adding behavioral modules, he may also be trying a phenomenological approach to some extent.

Chapter 7

Artificial Neural Networks and Machine Evolution

This chapter introduces artificial information networks that imitate the human nervous system. It also discusses the evolutionary development of machines using such networks and those in which the theory of evolution in biology is applied. If human consciousness or the mind consists of networks of nervous systems that connect the brain and the body, we should be able to imitate these human functions with networks of artificial nerves and using computer programs. It is only natural to believe that human consciousness and the mind came to be in the course of the phylogenetic development of humans. The latter part of this chapter describes possibility of a machine system autonomously acquiring something like human consciousness or the mind by evolving and developing artificial nervous systems, applying the principles of evolution theory and using computers.

7.1 Neural Networks

Braitenberg's vehicles behaved variously via signals directly transmitted from the sensors to the drive motors. His vehicles are a

Creation of a Conscious Robot: Mirror Image Cognition and Self-Awareness
Junichi Takeno
Copyright © 2013 Pan Stanford Publishing Pte. Ltd.
ISBN 978-981-4364-49-2 (Hardcover), 978-981-4364-50-8 (eBook)
www.panstanford.com

simple mechanical system with neural pathways for transmitting signals from sensors to drive motors. The electronic pathways are "neural pathways" mimicking those of a living organism. In other words, the behavior of the neural pathways determines the behavior of the vehicle.

If we could cleverly build neural pathways connecting sensors and drive motors in a robot, it would be able to perform various behaviors. Perhaps it would even be possible to create "consciousness" or "mind" on a computer or other machine systems using such artificial neural networks.

7.1.1 Hebb's Rule

Artificial neural pathways simulating the neural pathways of organisms are already known. They are generally called artificial neural networks, or simply neural networks. Neural networks were first introduced in 1949 by Canadian psychologist Donald Hebb, who proposed a theory known as Hebb's rule. Hebb's rule says that when two connected neurons are simultaneously activated, the strength, i.e., the weight, of the synapse between the two neurons increases. For example, when neurons x_1 and y_2 are simultaneously activated, the relevant synaptic weight w_{12} increases (Fig. 7.1). The synaptic weight is also simply called weight and is also known as the synaptic value. The synaptic weight is defined by the following equation:

$$\Delta w_{12} = x_1 \times y_2$$

or more generally:

$$\Delta w_{ij} = x_i \times y_j \tag{7.1}$$

where x_i and y_j are presynaptic and postsynaptic neurons, respectively (see Fig. 7.1).

Let the output $(y_1, y_2, y_3) = (1,0,0)$ be expected for the input pattern $(x_1, x_2, x_3, x_4) = (0,1,0,1)$ for this neural network. In this neural network satisfying the above input and output conditions, two pairs of neurons, x_2, y_1 and x_4, y_1, are activated according to Hebb's rule. Synaptic weight w_{21} between x_2 and y_1 is unity, i.e., the strength of the synapse between these two neurons increases. This condition also applies to w_{41} connecting x_4 and y_1 (Fig. 7.1(1)).

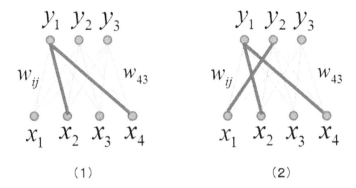

Figure 7.1. Application of Hebb's rule.

All other synaptic weights do not increase and are zero because, from Eq. 7.1, $w_{11} = 0$, for example, since $x_1 = 0$ and $y_1 = 1$. In Fig. 7.1, the synapses with an increased weight are shown using thicker lines. One lesson has been learned with Eq. 7.1.

Here is the second lesson. Let the output pattern $(y_1, y_2, y_3) = (0,1,0)$ be expected for the input pattern $(x_1, x_2, x_3, x_4) = (1,0,0,0)$. Then synaptic weight w_{12} increases in addition to the first lesson. The neural network after the second learning changes as shown in Fig. 7.1(2). As you might have already anticipated, the more lessons we learn, the more there will be thicker lines in the neural network until all the synapses have unity. Ultimately, all output patterns are ignited no matter what the input pattern is. The obvious error of this approach is that the synaptic weight increases with learning endlessly. The delta rule was devised to solve this problem.

7.1.2 *Single-Layer Neural Network and Delta Rule*

Let us consider the learning of a neural network as shown in Fig. 7.2(1). This type of neural network is called a single-layer neural network.

Let an input pattern ρ_n with n-input neurons be input into this neural network.

$$\rho_n = (x_1, x_2, x_3, \ldots, x_{n-2}, x_{n-1}, x_n)$$
$$= (1, 1, 0, 1, 0, 1, \ldots, 0, 1) \text{ (this is an example)}$$
where k is the total number of output neurons.

(1)

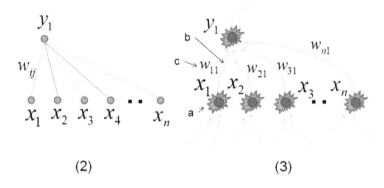

(2)   (3)

Figure 7.2.   Single-layer neural network and explanatory illustration.

$\Theta(M)$ is the linear output function to return a value continuously for variable M.

$$y_1 = \Theta(w_{11} \times x_1 + w_{21} \times x_2 + \cdots + w_{n1} \times x_n) = \Theta\left(\sum_{i=1}^{n} w_{i1} \times x_i\right)$$

$$y_2 = \Theta(w_{12} \times x_1 + w_{22} \times x_2 + \cdots + w_{n2} \times x_n) = \Theta\left(\sum_{i=1}^{n} w_{i2} \times x_i\right)$$

$$\vdots$$

$$y_k = \Theta(w_{1k} \times x_1 + w_{2k} \times x_2 + \cdots + w_{nk} \times x_n) = \Theta\left(\sum_{i=1}^{n} w_{ik} \times x_i\right)$$

$$y_{(j=1 \sim k)} = \Theta\left(\sum_{i=1}^{n} w_{ij} \times x_i\right) \tag{7.2}$$

This equation calculates the values of signals output from the neural network as the input patterns vary with the synaptic value w_{ij}. These complex equations are described in a more easily understood manner below.

Figure 7.2(2) is part of Fig. 7.2(1) that is extracted for the purpose of explanation. The extracted figure shows the relationship between output terminal y_1 and the input terminal pattern $(x_1, x_2, x_3, \ldots, x_n)$. Figure 7.2(3) is a network representation of the neurons in Fig. 7.2(2). The network consists of several neurons, a, as shown in Fig. 7.2(3), for example, and the neural pathways connecting the neurons, b, for example. Each neural pathway has a direction for transmitting signals, from x_1 to y_1, for example. Furthermore, each neural pathway has its own synaptic weight w_{ij} that indicates the strength of the flow of signals (c, or w_{11}, in Fig. 7.2(3), for example).

By modifying this weight, the input patterns can be changed to a desired output pattern. If, for example, weight w_{11} for information transmitted from input terminal x_1 to output terminal y_1 is zero, the signals received at x_1 cannot be transmitted to y_1. The value of output terminal y_1 in Fig. 7.2(3) is calculated by the first expression of Eq. 7.2. First of all, the input patterns are multiplied by their respective weight and the results are added to find the sum.

$$M_1 = (w_{11} \times x_1 + w_{21} \times x_2 + w_{31} \times x_3 + \cdots w_{n1} \times x_n)$$

Value M_1 represents the total weight-adjusted strength of the signals appearing on the input terminals. The value M_1 is then used to determine the value of output terminal y_1. The value of y_1 is determined only when value M_1 exceeds a certain threshold value.

Several methods are available to calculate the threshold value. Detailed explanations are omitted here, but in essence, the function $\Theta(M_1)$ is calculated.

The values of $y_2, y_3, y_4, \ldots, y_k$ are respectively determined by calculation. As a result, the output terminal patterns can be decided. All of these calculations are subsumed in the final expression of Eq. 7.2.

It is clear that various reactive systems can be constructed using these neural networks. The function of a system is determined by the structure of the network. Figure 7.2(1) shows one such neural network structure. The components of a "structure" here include the wiring connecting the terminals and the synaptic weight that determines the strength of the signals. Once the structure of the neural network is determined, its function is also determined. A reactive system can thus be constructed by determining wiring and synaptic weight. The objective of creating a reactive system is to obtain desired values or output patterns from variable inputs.

In a word, a system is a function. In that sense, neural networks may be said to be a computer program. Input a pattern into the system, and the system calculates and outputs the result for you. This is a computing system, but what is important is that this is a parallel computing system.

We use Eq. 7.2 to obtain y_1. This is pseudo-computer processing or sequential computing. Originally, all other output terminals y_2, y_3, y_4, ..., y_k were supposed to be computed simultaneously in parallel as shown in Fig. 7.2(3), and the computation of just a few steps would be enough to finish the calculation. The capability of parallel computation is one of the most important advantages of the human nervous system. Unfortunately, humans have not yet acquired a computation system to effectively make use of the merits of the parallel system; they have to simulate parallel calculation using currently available serial system computers.

Let us now study the problem of how to obtain the desired output pattern given certain input patterns. Let S, called the supervised signal, in Eq. 7.3, be the desired output pattern of a neural network that is given several input patterns ρ.

$$\delta_j^\rho = \left(S_j^\rho - y_j^\rho \right) \tag{7.3}$$

This equation finds the difference between the supervised signal and the current output pattern. The synaptic weight has not yet been learned so that an error δ inevitably exists. First, the approach is to determine the synaptic value to minimize the error δ. This δ is generally called the delta error.

How should we modify the synaptic weight? One possible solution is to use Eq. 7.4.

$$\Delta w_{ij}^\rho = \delta_j^\rho \times x_i^\rho \tag{7.4}$$

This means that input value x_i multiplied by the delta error (δ) is used as the value (Δw) to influence the synaptic weight.

The underlying idea is that the value to influence the synaptic weight is basically proportional to both the delta error and the input value. The value Δw can be a negative number, and therefore the problem experienced with a direct application of Hebb's rule is avoided.

The next step is to add Δw to each synaptic value in the neural network to obtain new values. When this operation is repeated for several training patterns, the synaptic weight converges to a certain value, and at that state we can say that learning is successful. Actually, we determine that convergence is reached when a very small value is observed.

Equation 7.5 is called the error function. It is one of the popular methods to determine convergence. This is generally known as the error of least squares. The error is intentionally squared for the purpose of always obtaining a positive number for the result and also for the purpose of excessively showing the estimated error.

$$E = \frac{1}{2} \sum_{\rho}^{m} \sum_{j}^{n} \left(S_j^\rho - y_j^\rho \right)^2 \tag{7.5}$$

Clearly, the delta rule is more powerful than Hebb's rule. Later, it was found that this system had a defect. The single-layer neural network, mentioned earlier, is incapable of an exclusive OR (XOR) operation of a logic circuit. The truth table of function $y = $ xor (x_1, x_2) is shown in Fig. 7.3(1). This function is a subtraction in a sense. The possibility of XOR processing was checked since XOR is an important operator in a computer. The input and output values are plotted on the two-dimensional coordinates as shown in Fig. 7.3 (2). As you see in the figure, the white and black dots cannot be separated by a straight line. The white and black dots represent 0 and 1 outputs, respectively, and the coordinates correspond to two input values.

In technical terms, we say that XOR is not linearly separable. There is no way you could draw a line that would separate the space on the coordinates.

Logical disjunction (OR), logical conjunction (AND), and logical negation (NOT) are linearly separable (Figs. 7.3(4), (6), and (8)), but XOR is not, which was pointed out as a problem.

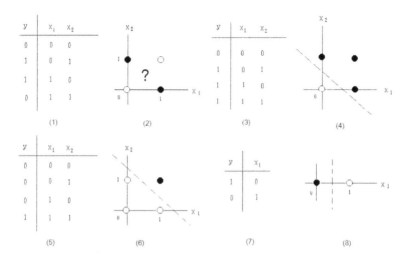

Figure 7.3. Linearly separable?

7.1.3 Feed-Forward Network and Back Propagation Method

The linear-separation problem was solved by Rumelhart *et al.* in 1986. They set up a second neural network above the first one as shown in Fig. 7.4. The result is a two-layered neural network. An intermediate (hidden) layer was provided between the input and output layers. The intermediate layer stores signals coming from the input layer temporarily and delivers them to the output layer after adjusting the signals.

Rumelhart *et al.* introduced nonlinear output functions for respective neurons to achieve nonlinear separation.

The sigmoid function (Fig. 7.5(1)), which is continuous, differentiable, and nonlinear and is frequently used as a nonlinear output function, is expressed by the following equation:

$$\Theta\left(x\right) = \frac{1}{1 + e^{-kx}}$$

Other available output functions include linear (Fig. 7.5(2)) and step functions (Fig. 7.5(3)). Output functions are like a filter in the sense that they give some features to the output values. The step function, for example, limits the output values to be either 0 or 1,

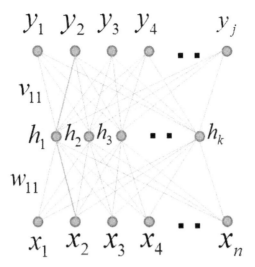

y_1 y_2 y_3 y_4 y_j

v_{11}

h_1 h_2 h_3 h_k

w_{11}

x_1 x_2 x_3 x_4 x_n

Figure 7.4. Two-layered neural network.

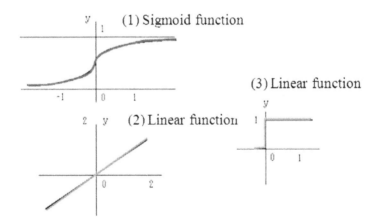

(1) Sigmoid function

(3) Linear function

(2) Linear function

Figure 7.5. Threshold functions.

while the sigmoid function outputs continuous values between 0 and 1.

How does a two-layered neural network learn a desired output pattern? The structure is somewhat complex, but basically the process performed by a single layer is performed twice. The basics of learning are the same as a single-layer network. The synaptic weights of the two layers together are adjusted such that the supervised signals S (desirable outputs) are output in response to input patterns ρ_n. Initially, a near-zero random number is selected for the synaptic value. An input pattern for training t_i is given to input terminal x_i (Fig. 7.4; Eq. 7.6).

Then we calculate value h_k^ρ of the intermediate layer (Eq. 7.7). This calculation uses Eq. 2.2, without any modification, which calculates the output values for a single-layer neural network, except that the input terminal is x_i and the output terminal (intermediate terminal, actually) is h_k^ρ. Output value y_j^ρ is then calculated (Eq. 7.8). This calculation also uses Eq. 2.2, without any modification, which calculates the output values for a single-layer neural network, except that the input terminal (intermediate terminal, actually) is h_k^ρ and the output terminal is y_j.

Note, however, that there are two kinds of delta errors in a two-layer network. The delta error of the second layer δ_j^ρ is first calculated, and then using this value, the delta value of the first layer δ_k^ρ is calculated.

We first calculate the generalized delta error δ_j^ρ for each output value. S_j^ρ is the desired supervised signal to be output (Eq. 7.9). Then we calculate the generalized delta error δ_k^ρ for each value of the intermediate layer (Eq. 7.10).

$$x_i^\rho = t_i^\rho \tag{7.6}$$

$$h_k^\rho = \overset{*}{\Theta}\left(\sum_{i=1} w_{ik} \times x_i^\rho\right) \tag{7.7}$$

where $\overset{*}{\Theta}$ is a sigmoid function.

$$y_j^\rho = \overset{*}{\Theta}\left(\sum_{k=1} v_{kj} \times h_k^\rho\right) \tag{7.8}$$

$$\delta_j^\rho = \overset{*}{\Theta}\left(\sum_{k=1} v_{kj} \times h_k^\rho\right) \times (S_j^\rho - y_j^\rho)$$
$$= y_j^\rho \times (S_j^\rho - y_j^\rho) \tag{7.9}$$

$$\delta_k^\rho = \overset{*}{\Theta} \left(\sum_{i=1} w_{ik} \times x_i^\rho \right) \times \sum_j v_{jk} \times \delta_j^\rho$$

$$= h_k^\rho \times \sum_j y_{jk} \times \delta_j^\rho \qquad (7.10)$$

In Eq. 7.10, delta error δ_j^ρ of the output layer is multiplied by the synaptic value of the second layer in reverse. The result is then multiplied by the respective values of the intermediate layer that were calculated before to derive delta error δ_k^ρ of the intermediate layer. This procedure is called back-propagation because the error is returned from the second to the first layer. We then calculate the modifier values (Eq. 7.11) to modify the postsynaptic and presynaptic values. The postsynaptic and presynaptic values are modified by Eq. 7.12. This equation corresponds to Eq. 7.4 used for a single-layer network. This is an estimation of the difference in synaptic weight between the second and first layers.

$$\Delta v_{jk}^\rho = \delta_j^\rho \times h_k^\rho$$

$$\Delta w_{ki}^\rho = \delta_k^\rho \times x_i^\rho \qquad (7.11)$$

$$\overset{(t)}{v}_{jk} = \overset{(t-1)}{v}_{jk} + \eta \Delta v_{jk}^\rho$$

$$\overset{(t)}{w}_{ki} = \overset{(t-1)}{w}_{ki} + \eta \Delta w_{ki}^\rho, \text{ where } 0 < \eta < 1 \qquad (7.12)$$

Equation 7.12 calculates changes to be given to the synaptic weights of the first and second layers. In this equation, η is called the learning rate. As η approaches unity, the effect of learning increases. The network repeats the calculation according to the above procedure using several training sets until the difference from S reaches the minimal level. Convergence is generally determined by the following evaluation function:

$$E_B = \frac{1}{m} \sum_\rho^m \left(\frac{1}{n} \sum_j^n (s_j^\rho - y_j^\rho)^2 \right) \qquad (7.13)$$

where m and n are the total numbers of the training sets and output terminals, respectively. Learning is determined to have completed when this equation approaches zero, which means that y_j (the value

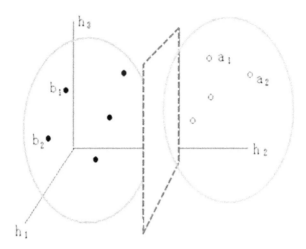

Figure 7.6. Split of activated vector space and categorization of activated vectors.

of the output terminal) is sufficiently close to S_j (the supervised signal). This type of network is called a feed-forward network.

Each terminal of the intermediate layer can be considered to be a vector with a unique coordinate system in a spatial representation of multiple dimensions (i.e., multiple terminals). This type of vector is called an activated vector, and the space where it is observed is called the activated vector space. The coordinate values are a function of the values of the input and output terminals. For this reason, various input patterns can be categorized in the activated vector space.

If, for example, the activated vector space (Fig. 7.6) of an intermediate layer is separable into black (b_i) and white (a_i) dots, we can categorize the vectors into two groups.

7.1.4 *Recurrent Neural Networks and Their Functions*

"Recurrent" means that the values of the output terminals are returned to the intermediate terminals. An example is illustrated in Fig. 7.7. In this example, value y_j of the output terminal is input into intermediate terminal h_k in the next time step. The role of a recurrent neural network is to dynamically change the outputs as the input values change with time.

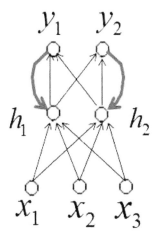

Figure 7.7. Image of a recurrent neural network.

Conversely, the memory of reactions continuing from the past to the present that have brought about the current output still exists. A detailed explanation follows referring to Fig. 7.7.

Figure 7.8 shows how outputs change with time. The output of this network at $t = n$ is (y_{n1}, y_{n2}). Note that in addition to the input pattern (x_{n1}, x_{n2}, x_{n3}), the output pattern $(y_{(n-1)1}, y_{(n-1)2})$ at $t = n - 1$ is also input into the intermediate layer at $t = n$.

The value of an output terminal is determined by the values of the intermediate terminals. The behavior of the values of the

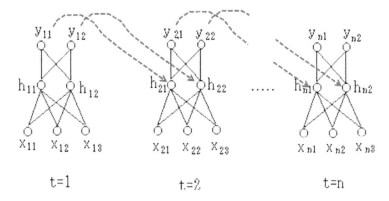

Figure 7.8. Dynamic reaction of a recurrent neural network.

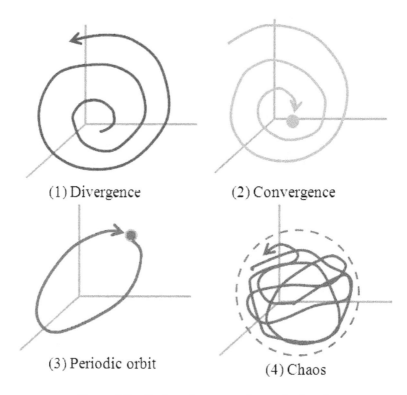

(1) Divergence   (2) Convergence

(3) Periodic orbit   (4) Chaos

Figure 7.9.   Motion of a recurrent neural network.

intermediate terminals is shown in the activated vector space in Fig. 7.9.

The activated vector can take on one of four different states according to the change in input patterns: (1) divergence, (2) convergence, (3) periodic orbit, or (4) chaos. In any of these states, the vector values vary with time, allowing us to build a dynamic system. States 2 and 3 are easy to use, and both can be used for building a dynamic reaction system. State 3, in particular, relates to periodic action. It is possible to move the hands and feet of a robot using this vector. State 4, chaos, is related to unpredictable motion within a certain boundary. This is interesting when you remember that the human body has been called a mass of chaos. It is said, for example, that the heartbeat is related to chaos.

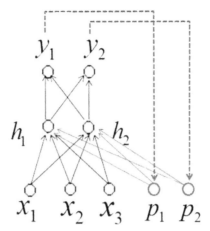

Figure 7.10. Example of an actual recurrent neural network.

There is a convenient method for creating a recurrent network. Referring to Fig. 7.10, prepare recurrent terminals (p_1, p_2) beforehand to which you will later copy recurrent values. When the output pattern (y_1, y_2) is determined, copy its values to the recurrent terminals simultaneously.

7.1.5 *Summary and Observations*

Artificial neural networks have been briefly explained in this section. Specifically, feed-forward and recurrent networks are discussed. In a feed-forward network, an output pattern is uniquely determined by the input pattern. This type of network is capable of learning basic logic functions and thus, in principle, can achieve all logic functions in a network.

We have also shown that input patterns of a feed-forward network can be categorized by representing the intermediate layer of the network in the activated vector space. Recurrent networks are capable of learning dynamic output states that change with time. The activated vector space of the intermediate layer of a recurrent network has four output states: divergence, convergence, periodic orbit, and chaos. Their respective learning behavior and usability have been described.

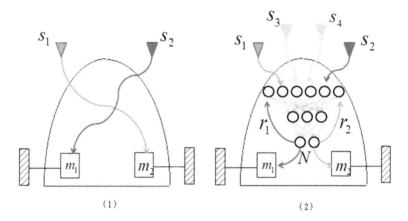

Figure 7.11. Braitenberg robot and its development.

7.2 Theory of Robot Evolution

In the previous chapter, we explained that neural networks can learn output patterns in response to input patterns. As such, it would be possible to build a machine with sensors and motors or other drive units, in which the sensors generated input patterns and the resultant output patterns were connected to the motors, to make the machine move in response to the values of the sensors. This was actually done by Braitenberg. How the machine reacts is determined by the neural network connected between the sensors and the motors.

In the Braitenberg robot, the sensors and motors are connected by simple circuits. With such a simple design, the robot achieved remarkable motion performance (Fig. 7.11(1)).

We are now able to replace the simple circuit with a complex neural network in the Braitenberg robot (Fig. 7.11(2)).

It has gradually become known that it is not so easy for a robot to learn complex neural networks, in particular, a recurrent network. Let's stop and think this over for a moment. The original objective of learning a neural network is to obtain desirable output patterns for a given set of input patterns. To obtain desirable output patterns, we only need to adequately modify the synaptic values in the neural network.

7.2.1 *Machine Evolution Approach*

Robotics researchers hit upon the idea of using the techniques of evolution that were broadly adopted in biological studies. They decided to apply the principle of the survival of the fittest in the development of robots.

Assume 10 robots decide their actions using their own respectively different artificial neural networks. Four of them perform desirable actions, while the others fail to do so (e.g., could not avoid obstacles). The four robots are determined to be the Fittest that survive. The surviving four robots create the next generation of robots. The principle of the survival of the fittest is again applied to the next generation, and robots that act desirably survive. If this process were repeated over some generations, we would eventually have robots that acted excellently. By simulating the principles of the evolution of organisms in this way, researchers created the machine evolution approach.

The robot shown in Fig. 7.12(1) has four sensors to generate input patterns. These input pattern values are transmitted to the output patterns (motors) via neural networks (Fig. 7.12(2)). The neural networks can be variously configured by how their junction of neurons and synaptic values are set. The variety of neural networks that can be configured means that the robots' behaviors can also be diverse. What I mean is that we only need to adequately vary the artificial neural networks for the actions we wish the robot to perform.

The neural network shown in Fig. 7.12(2) has 17 neurons connected by 27 synapses. The state of connection is shown in the matrix in Fig. 7.12(3). The top row and the leftmost column denote all neurons ($n_1, n_2, n_3, \ldots, n_{17}$). The element w_{12}, for example, in the matrix represents the synaptic value from neuron n_1 to n_2. Generally, the element w_{ij} in the matrix represents the synaptic value from neuron n_i to n_j. This matrix thus shows all of the synapses and synaptic values for the entire set of neurons. Let us consider the simple case of three neurons n_1, n_2, and n_3. The area enclosed by the dotted lines in Fig. 7.12(3) is the region of interest. This region is graphically represented in Fig. 7.12(4). In this diagram, the three neurons and all of their synapses with the corresponding values are

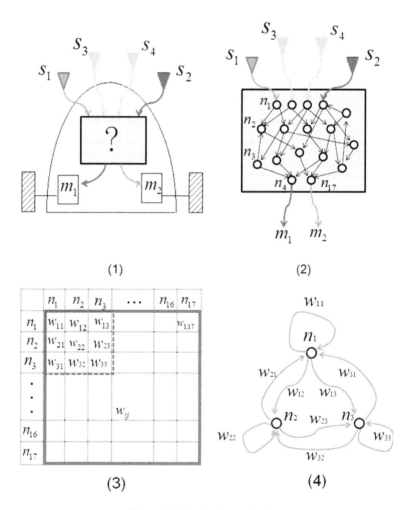

(1)   (2)

(3)   (4)

Figure 7.12.   Robot evolution.

covered. In discrete mathematics, this diagram is called a directed complete graph with three nodes. Synaptic weight w_{11} is the neural pathway from neuron n_1 to itself, which is called a self-loop in technical terms and is a kind of recurrent pathway.

Since the elements in this matrix determine the behavior of a robot, all of the behaviors of the robot can be programmed by specifying all combinations of the elements.

As such, in a direct way of thinking, it would seem to be possible to discover neural networks that constitute the substance of human consciousness or the mind in the combinations of elements of all of the synapses. If the neural network constituting human consciousness or the mind is a redundant system, however, I am afraid the possibility of discovery is very low.

Putting aside the difficult problem of consciousness and the mind for now, it is clear that the optimum behavior for the robot can be selected from among these combinations of elements. For three neurons, the maximum number of neural pathways (number of synapses) is 9 (number of neurons × number of neurons). In Fig. 7.12(4), the maximum number of neural pathways is 289 since there are 17 neurons.

Let the maximum value that a synaptic weight w_{ij} can take be a one-digit number in the decimal system, i.e., an integer between 0 and 9. Then in the case of the 3 neurons shown in Fig. 7.12(4), each synaptic value can take 10 different numbers, and there would be 1000 (10^3) different pattern possibilities in this neural circuitry. When we run the robot by applying each and every possible pattern to its neural circuitry shown in Fig. 7.12(4), we can find the optimum pattern. In the case of the 17 neurons shown in Fig. 7.12(3), the total number of possible patterns reaches 100000000000000000 (10^{17}), which practically prohibits computation, although it is theoretically possible. In technical terms, this solution brings about an exponential explosion of computation time. In the present example, the synaptic values are limited to one-digit decimal integers. If this limitation were removed and any real number could be used, there would have been no possibility of computation.

To avoid this problem, we do not check all possibilities one by one, but instead we check areas with higher possibilities of finding the solution sequentially until a suboptimal solution is identified. A suboptimal solution is a practically acceptable solution that would be theoretically close to, though is not itself, an optimal solution. This approach was devised in the study of artificial intelligence and adopts the idea of biological evolution. Hence, the method is named "machine evolution."

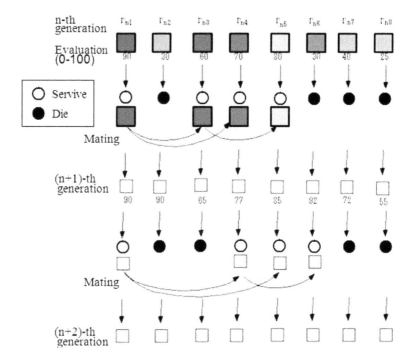

Figure 7.13. Biological evolution approach.

One of the popular machine evolution approaches is introduced now. Assume that eight robots survive in the n-th generation (Fig. 7.13). We evaluate the fitness function of each robot with respect to its environment on a 100-point scale.

The fitness functions are assigned by humans. For example, the evaluation mark would increase the faster a robot runs. The survival-of-the-fittest principle is applied to the robots after evaluation. Various discussions exist regarding how many marks would qualify for the fittest. We decide to select "die" for four robots with poor evaluations, and the remaining four robots survive as the fittest (selection). These four surviving robots, with more than 60 points as the evaluation mark, are mated. Various mating methods are conceivable. Only the surviving robots are mated or some dead ones may be recovered temporarily for mating purposes. The mating problem will be discussed in detail later.

For now, the four surviving robots reproduce eight robots by mating. These eight robots are the $(n + 1)$th generation. The robots of this new generation are evaluated to select the surviving robots. At this time, the value of the fitness function increased to more than 77. We can hopefully expect the fitness function to increase gradually by repeating the above process generation after generation. Repetition stops when robots of a high evaluation are found. This is a breakthrough approach in that the details of the artificial neural network are omitted and the overall picture is grasped. This sounds like saying that good results are important no matter how the artificial neural network is structured. That is why this approach is enthusiastically hailed by behaviorists.

The machine evolution approach is said to have been initiated by John Holland and L.J. Fogel, both in the United States, and Ingo Rechenberg in Germany, in the 1960s almost simultaneously (Pfeifer and Sceider, 2001). Holland advocated a genetic algorithm, while Fogel promoted evolutionary programming, and Rechenberg developed an evolutionary strategy technique. The three researchers seem to have been working in different fields of study, but in reality there is no significant difference among them from the broader perspective of machine evolution.

The machine evolution approach has made a great progress historically hand in hand with the study of artificial life.

"Artificial life," as defined by Christopher Langton of the Santa Fe Institute, "is an artificial system that behaves like a living organism." In the study of artificial life, artificial living organisms, called creatures, reproduce themselves, grow, prosper, decay, and die by simulation in computers.

In their studies, artificial life researchers naturally discuss the evolution of their artificial living organisms. However, their research is performed in an ideal environment of computers, which is markedly different from the environment of machine evolution research.

Researchers of machine evolution claim that machines acquire knowledge through interaction with the natural environment. This very belief prevents them from attempting to build a natural environment on a computer, which remains a decisive drawback.

I personally believe that the study of artificial life and Braitenberg's robot studies were fused to give rise to the study of machine evolution and spurred its rapid development. In other words, the evolutionary technique of artificial life and Braitenberg's robots that moved around in an actual environment have merged. Let me explain in more detail, beginning with genotype and phenotype.

Humans have genes. The genes are used as a plan for producing and developing the body and other parts of a human. Genes are a plan and the human body is a representation of the plan. Using this analogy, we may say that in machine evolution, a genotype corresponds to a numerical sequence or a character string, and phenotype to a machine system created from or representing the numerical sequence or character string (Fig. 7.14(1)). The machine system as a phenotype may be a hexapod centipede or a function unit that generates adequate output patterns in response to input patterns.

In the discussion to follow, we consider the function unit of Braitenberg's robot shown in Fig. 7.12(1). The component with the question mark (?) in the figure is the function unit.

Our robot has four sensors S_1, S_2, S_3, and S_4. When an external stimulus is given to the sensors, the function unit responds and outputs signals to two drive motors m_1 and m_2. The behavior expected for the robot is to "continue moving without collision as much as possible." We assume that the function unit incorporates a single-layer neural network of four inputs and two outputs (Fig. 7.14(1)). The structure of the network itself could be the result of evolution, but we select a simple scenario for now.

We then determine genotypes and use four-digit binary sequences because of the ease of representation in genes. Since our robot changes its behavior according to the synaptic weight, we relate the genotype to the synaptic weight. We further assign values to respective synaptic weights, which are the elements of the genes in our example. The values are determined by generating random numbers. The synaptic weights are four-digit binary numbers, which are equivalent to 0 through 15 in the decimal system.

In our present study, we determine the synaptic weights as follows:

$$(w_{11}, w_{12}, w_{21}, w_{22}, w_{31}, w_{32}, w_{41}, w_{42}) = (7, 1, 5, 2, 3, 8, 1, 5)$$

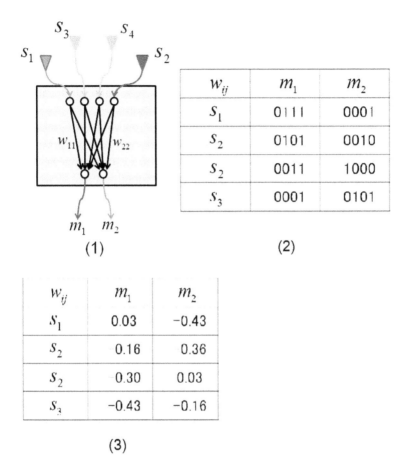

Figure 7.14. Genotype and phenotype of a machine.

The function unit has eight synapses, each having a four-digit synaptic value, i.e., the total number of bits of the genes of the function unit is 32 (8 × 4). The respective synaptic values are therefore represented in four bits as follows (Fig. 7.14 (2)):

$$w_{ij} = (0111\ 0001\ 0101\ 0010\ 0011\ 1000\ 0001\ 0101)$$

The genes of this function unit are now finally determined as follows:

$$g = (01110001010100100011100000010101) \qquad (7.14)$$

This is the genotype of our robot. Let us now calculate the phenotype for this genotype. Each synaptic value is divided by 15, and 0.5 is subtracted from the result. This gives you the phenotype (Fig. 7.14(3)). We divide by 15 to convert the synaptic values of 0 to 15 to values between 0 and 1.0. We subtract 0.5 to further convert the resultant values to those between −0.5 and 0.5. Negative numbers are intentionally introduced because inhibition values must be considered for certain synaptic junctions.

We have thus converted the synaptic values to values suitable for use in neural networks. It is assumed in our example that the input and output values are real numbers, including negatives, and therefore we use the values of the sigmoid function $\Theta(x)$ to which 0.5 is added as the threshold function of the neuro unit. This is necessary, for example, to run the drive motors in reverse. Merits include the fact that the sigmoid function is effective for inputs between −1.0 and 1.0, and that values near zero are suitable for the initial value of genes (Fig. 7.5 (1)).

It is somewhat difficult to understand that Fig. 7.14 (3) shows the phenotype. The elements in the phenotype directly indicate the synaptic values of the single-layer neural network built in the function unit, and the actual robot (Fig. 7.14 (1)) behaves on the basis of these synaptic values. As the phenotype is decided, the robot's behavior is also decided.

An explanation of population now follows. The genotype defined by Eq. 7.14 is an initial value. Let us prepare, in our example, 20 genotypes as the initial value. Random numbers may be used. When considering these 20 genotypes to be phenotypes, each behaves differently. These 20 genotypes are therefore phenotypes of 20 individuals at the same time. These 20 individuals constitute the population of the first generation.

Now we proceed to the selection process. We select individuals with a high fitness function from among these 20 individuals. In our example, robots capable of avoiding collision as much as possible are determined to have a high fitness function. One possible evaluation method would be to have each individual robot run and count the number of collisions observed within a unit time. Assume we select 10 individuals.

Various methods are available to select 10 individuals. One of the popular methods is roulette wheel sampling. Individuals are

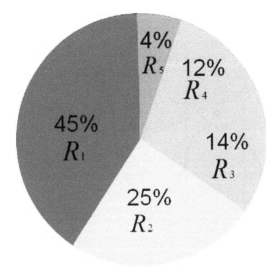

Figure 7.15. Roulette wheel sampling.

selected by probability R_i corresponding to relative fitness within the given population (Fig. 7.15).

The widths of the regions on the roulette wheel are proportional to the level of fitness. We spin the roulette wheel. When the pointer stops at region R_3, for example, the individual with that relevant fitness is selected. If there is more than one individual, we select one at random. Repeat this procedure 10 times, and you have selected 10 individuals. The advantage of the roulette wheel sampling is that individuals with various levels of fitness are selected with a certain balance for the given generation. "Balance" here means that individuals of not only a high fitness but also those of a middle and low fitness are selected as well. This method helps us approach the optimum solution, or, in technical terms, avoids jumping at the local maximum.

Assume the solution space for the distribution of genotype solutions is as shown in Fig. 7.16.

The solution distribution in our example has two different modes, each with a peak. This is a type of multimodal solution distribution. Peak p_1 has a low fitness value compared with the

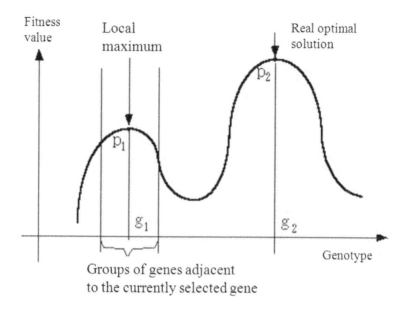

Figure 7.16. Solution space and multimodal solutions.

second peak p_2. Peak p_1 is the local maximum and peak p_2 is the optimum fitness. Genotype g_2 is therefore the desirable answer.

If you always selected individuals of a high fitness in this selection process, all of them could exist around genotype g_1. If this were the case, you would have ended up selecting individuals near the local maximum, and could get out of the local area, eventually failing to select genotypes that would surpass the local maximum.

Note that a solution space like the one shown in Fig. 7.16 is totally unknown to humans initially. Indeed, if it were known, we need not have any trouble.

It is known that the machine evolution approach is not effective unless the solution space is continuous and differentiable. In other words, for the machine evolution approach to work effectively, the fitness determined by certain genotype g_1 must not change abruptly (should be differentiable) from the fitness of the gene adjacent to g_1. Conversely, for a problem involving any sort of solution space, the machine evolution approach finds an answer somehow. This is the great point of the machine evolution approach.

Other methods for selecting individuals include elitism (to leave behind individuals with optimum fitness values in a generation), rank selection, and tournament selection. Rank selection refers to the use of ranking of individuals' fitness values, while tournament selection picks two individuals at a time for comparison for survival. After selection of individuals comes the process of reproduction.

We have earlier selected 10 out of 20 individuals. Genetic operations are performed on the selected individuals to reproduce 20 individuals again.

Two popular genetic operations are crossing and mutation. Crossing refers to the crossover of genes of the selected individuals. Assume we pick two genes g_1 and g_2 from a selected individual.

$$g_1 = (0111\ 0001\ 0101\ 0010\ 0011\ 1000\ 0001\ 0101)$$
$$= (g_{11} \qquad\qquad g_{12} \qquad\qquad)$$
$$g_2 = (0011\ 1000\ 0001\ 0101\ 0111\ 0001\ 0101\ 0010)$$
$$= (g_{21} \qquad\qquad g_{22} \qquad\qquad)$$

Crossing occurs between the sixth and seventh characters from left in the above gene sequences. Random numbers may be used. Two crossed genes g'_1 and g'_2 are generated.

$$g'_1 = (0111\ 0000\ 0001\ 0101\ 0111\ 0001\ 0101\ 0010)$$
$$= (g_{11} \qquad\qquad g_{22} \qquad\qquad)$$
$$g'_2 = (0011\ 1001\ 0101\ 0010\ 0011\ 1000\ 0001\ 0101)$$
$$= (g_{21} \qquad\qquad g_{12} \qquad\qquad)$$

The original genes $g_1 = (g_{11}g_{12})$ and $g_2 = (g_{21}g_{22})$ turned out two new genes $g'_1 = (g_{11}g_{22})$ and $g'_2 = (g_{21}g_{12})$ by crossing. The phenotype of the gene has changed as follows:

$$g_1 = (w_{12}, w_{12}, w_{21}, w_{22}, w_{31}, w_{32}, w_{41}, w_{42})$$
$$= (7, 1, 5, 2, 3, 8, 1, 5)$$

to

$$g'_1 = (7, 0, 1, 5, 7, 1, 5, 2)$$

The above process applies to g_2 in like manner.

Mutation is described now. Mutation means that part of the information of the genes of a selected individual is overwritten.

One element of the gene is subjected to bit conversion, or 1 and 0 are overwritten by 0 and 1, respectively. The position of the bit conversion can be arbitrarily determined.

$$g_1' = (0111\ 0000\ 0001\ 0101\ 0\underline{1}11\ 0001\ 0101\ 0010)$$

$$\uparrow \text{Position of mutation}$$

$$g_1'' = (0111\ 0000\ 0001\ 0101\ 0011\ 0001\ 0101\ 0010)$$
$$= (7, 0, 1, 5, 3, 1, 5, 2)$$

In the above example, the value of the 18th position (underlined 1) of gene g_1' changes by mutation and a new gene g_1'' is created. These gene-operated genes are reproduced to give birth to the next generation. The procedure is repeated until genes of a desirable fitness are found.

It is understood that the machine evolution approach can not only be applied to the design of the above-mentioned function unit but also be expanded to the study of the shape of the machine systems in which the function unit operates.

Karl Sims is a famous researcher on the evolution of shape. His studies are very interesting, and I invite the readers to study his work.

7.2.2 Summary and Observations

This section explained the machine evolution approach. The machine evolution approach simulates the biological evolution method called the survival of the fittest. New generations are created by mating, and the fittest among the individuals is selected for further mating. This biological method is adopted in the machine evolution approach using an engineering technique. "Fittest" means that one is adapted to the environment. Our eventual goal is the survival of excellent generations. I also explained that this approach is a powerful method for complex neural networks to learn successfully. This approach is useful in that it can find specific solutions to problems for which humans are utterly at a loss as to how to find solutions.

In the machine evolution approach, humans specify the genotypes, set pairing of genotypes and their phenotypes, and stipulate the fitness function based on the survival-of-the-fittest principle. All of the processes of evolution are then left to computers. It is naturally possible to set genes as the object of evolution. As a result, artificial living organisms beyond the imagination of humans could appear. This phenomenon is called emergence.

There are problems, though.

In the machine evolution approach, evolution takes place in interaction with the environment in principle, but it takes an incredible amount of time to observe how and to what extent the evolved individuals will satisfy the fitness function. The machine evolution approach is used jointly with computer simulators and with back propagation and other neural network learning methods to shorten the computation time. Despite all of these efforts, no essential solution has yet been found. Furthermore, no one has ever clarified theoretically the reason why the machine evolution approach identifies good answers swiftly.

Chapter 8

Machine Consciousness

I have described the achievements of general physiological and physical research on the human brain and also introduced outstanding historical research studies on human consciousness and the mind and the relevant achievements. There is a long history of research encompassing psychology, philosophy, cerebral physiology, information science, and cognitive science.

Braitenberg, a cybernetics researcher, demonstrated the possibility of constructing something resembling the human mind on a machine.

I explained mathematical methods for understanding and using artificial neural networks that are potentially capable of artificially reproducing human neurons and nerve networks. Roboticist Brooks proposed subsumption architecture — the world's first concept of its kind. He created a robot that moved around at all times. The behaviors of the robot were hierarchized and higher-level behaviors were achieved as the robot built up a behavior hierarchy pyramid, eventually fulfilling the emergence of consciousness and the mind, according to Brooks.

Brooks further outlined the techniques of mechanical evolution in which the principle of the survival of the fittest in biological evolution is applied using an engineering approach. On the basis of

Creation of a Conscious Robot: Mirror Image Cognition and Self-Awareness
Junichi Takeno
Copyright © 2013 Pan Stanford Publishing Pte. Ltd.
ISBN 978-981-4364-49-2 (Hardcover), 978-981-4364-50-8 (eBook)
www.panstanford.com

this knowledge, I will introduce some studies closely related to the development of a conscious robot.

8.1 Walter's Turtle

W. Grey Walter (1910–1977), a US-born researcher, will be remembered for his world's first life-simulating robot. Between 1948 and 1949, he created two robots named *Elsie* and *Elmer*. This was the era when the ENIAC electronic computer had just been developed. Driven by batteries, Walter's robots travel on three wheels. An oblong turtle-shaped shell protects the body, and a small chimney-like shape protrudes from the shell (Fig. 8.1a,b). The front steering wheel turns and determines the traveling direction of the robot. The two rear-drive wheels control forward and backward travel. A light sensor (a photoelectric cell) is mounted on top of the chimney to observe the environment. The chimney is directly coupled to the steering wheel, so that when the sensor turns toward a light, the

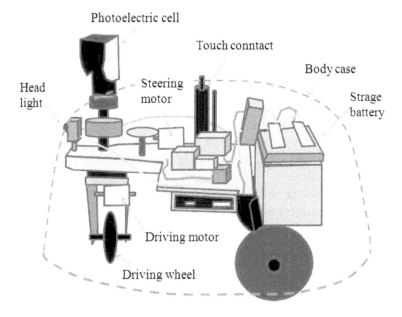

Figure 8.1. Internal structure of the robot.

steering wheel turns in the same direction. The robot is provided with a touch sensor to detect a collision with an obstacle. The touch sensor is mounted on the shell and a collision is detected when the shell hits an obstacle while traveling. The robot has a headlight on its front to show its position. The intensity of this lamp seems to be relatively low. It is said that this lamp also serves to confirm the energization of the robot (a pilot lamp). The robot's electronic circuit comprises two electron tubes, several relays, and switches. This is a very simple circuit compared with a modern computer. Of course, this circuit was analog not digital.

Several landmark experiments performed by Walter are introduced in this chapter. His experiments were performed in a dark room since the robot was designed to react to light.

The first experiment that I would like to introduce was conducted with a high-intensity lamp and with a relatively large obstacle placed in the room. The robot was positioned such that it could not see the light due to the obstacle. The robot initially moved around without purpose. Accidentally, the robot hit the obstacle and moved back. As it moved back, the robot slightly changed its traveling direction (because it was designed to do so). The robot stopped and then again moved forward. After repeating this series of movements for some time, the robot accidentally cleared the obstacle and the lamp was visible to the photosensor of the robot. The robot reacted and turned toward the light and advanced. The robot seems to be designed to stop when the light intensity reached a certain level, then move back a little, and thereafter resume forward motion. When viewed from a macroscopic perspective, what the robot did was to go around, not toward the lamp. This behavior reportedly looked like a living being that was aiming at its prey while making a detour to avoid obstacles.

In the next experiment, Walter uses two robots and puts away the large lamp that he used in the previous experiment. The two opposing robots move ahead toward each other, aiming at their respective headlamps. Since the light intensity of the headlamps was small, the two robots repeatedly collided and backed up. The robots were described to be "engaging in a conversation."

Another feature of the robots is that they returned to their home base, called a dock, when their batteries were low for recharging. A strong light installed in the dock guides the robot into the hanger.

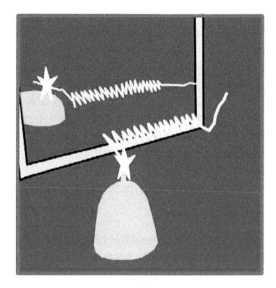

Figure 8.2. Walter's self-recognition experiments.

Walter's robot could move almost perpetually by recharging its energy like a living being.

The last noteworthy experiment of Walter that I would like to introduce is his self-recognition experiment using a mirror (Fig. 8.2). A robot is placed in front of a relatively large flat mirror. Apparently, this is similar to the experiment with two robots introduced above except that the second robot is the image in the mirror. The robot reacts to the image of the flame of a candle set on its shell and moves toward the image. The robot hits the mirror and thereafter repeatedly backs up, moves forward, and collides with the mirror. How is this different from the previous experiment? In the previous experiment, the movement of the two robots gets disordered as they repeat the same movement, but in this mirror experiment, the movement of the "two robots" (i.e., the robot and its reflected image) is synchronized because the mirror is used as an intermediary. Because of its intrinsic property, the mirror traces the motion of the robot moving in front of it.

I think this was the world's first experiment that showed the possibility of having a robot discriminate between its self-image in the mirror and another robot. It is a wonder that such an experiment was successfully conducted in as early as 1950. This experiment suggests the possibility of mirror image cognition by robots. Mirror image cognition by a robot is a theme of study aimed at unraveling the mystery of human consciousness by exploring the human mechanism of mirror image cognition through the development of a robot with a built-in mechanism of awareness of its self-image reflected in a mirror.

A question, however, remains: Does Walter's robot really discriminate between its self-image and the other? This experiment seems only to show that the robot can discriminate between the other and some other. Can we simply conclude that "some other" is none other than the self? I think it is difficult to judge that "some other" is the robot's "self-image."

I personally believe that in this experiment the robot discriminated between the other and some other, or between two different others. In order to assert that some other is the robot's self-image, some concrete gadget to scientifically show their relevance must be built into the recognition mechanism of the robot, or at least, the relevance must be shown with scientific proof.

8.1.1 *Summary and Observations*

The small robot called a turtle developed by Walter was the world's first biological robot. Walter conducted many interesting experiments and demonstrated the possibilities of robots to people worldwide at expositions. Prof. Braitenberg and many other researchers owe their various enlightenments to Walter's experiments. Walter's experiments using mirrors, in particular, were epoch-making in the sense that they were the first challenge to the difficult problem of self-image cognition. I wish to acknowledge that his study was the first step toward the success of mirror image cognition by robots, and I express my respect for his achievements.

8.2 Kitamura's Robot

The conscious robot developed by Tadashi Kitamura seems to be deeply influenced by Brooks' subsumption architecture. Kitamura's originality is best exhibited by the fact that he exquisitely correlated the hierarchy of consciousness devised by Vietnamese philosopher Tran Duc Thao (1917–1993) to the hierarchization of behaviors formulated by Brooks using the Expression Model for Consciousness and Behavior that he revised and compiled (Kitamura, 2000).

The Expression Model for Consciousness and Behavior advocated by Kitamura is explained as follows. He sets nine conscious levels. Like Brooks' subsumption architecture, any conscious level subsumes all the underlying levels.

Level 0 is the lowest level, and from the viewpoint of phylogeny, Level 0 is equivalent to the basic knowledge of protozoa in the waking state. The behavior corresponding to this level is the basic reactions needed for existence. Levels 4 and 5 are nearly equivalent to those of mammals. Consciousness in these levels includes the stable discerning of objects and stable and sustainable feelings as well as the temporal and spatial relationships between objects. The relevant behaviors are detour, search, aim, be prepared, control limbs, and use media. Levels 7 and 8 correspond to those of humans. Consciousness at levels 7 and 8 include symbolic representations and concepts, and the relevant behaviors are the manufacturing of tools and the use of language.

The basic idea behind the hierarchization of the consciousness levels is the belief that the evolution of the brain nerves pertaining to behavior control is deeply related to the phylogenetic evolution of living organisms. On the basis of Thao's theory, Kitamura estimated the consciousness development process from the development process of the body and behavior considering phylogenetic evolution (Kitamura, 2000). Kitamura then tried to build his model in a machine. Figure 8.3 shows the conscious robot designed by Kitamura. The conscious levels range from 0 to 5 in this figure. The conscious system is basically designed to evoke consciousness by using the values of sensors to select the corresponding behavior, and then drive the motors according to the selected behavior using a drive system. Figure 8.3 also suggests the similarity of Kitamura's idea to Brooks' subsumption architecture. Of course, Kitamura's architecture is significantly different from Brooks' theory in that Kitamura clearly defines consciousness to be possessed by a robot.

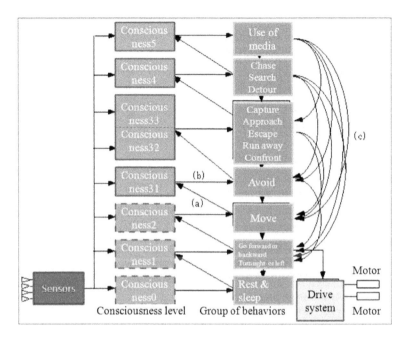

Figure 8.3. Kitamura's conscious robot model (partly modified by the author).

The relationship between consciousness and behavior is described now. Assume a robot in which a behavior is suppressed while the robot is performing that behavior. The robot moves to the next higher conscious level, selects and engages in the corresponding behavior. For example, a robot is at conscious level 2 and is performing a "move" behavior. The robot's "move" behavior is now suppressed. At this time, the robot raises its conscious level by one, or from level 2 to 31 (Fig. 8.3a). Specifically, the robot moves from the conscious level of the sensorial area evaluated by good and bad to another conscious level of temporary feelings such as emotions. This transition is effected by selecting the "avoidance" behavior (Fig. 8.3b). This series of behaviors can be summarized as "the robot avoided because its movement was blocked." The flow of consciousness would be that the robot avoided because it "was made sad to have met something bad."

Kitamura believes that the suppression of behavior is the suppression of consciousness. This is because he believes that Merleau-Ponty's assertion that "consciousness means behavior" is justifiable. This is a difficult phrase to understand. To put it plainly, the behavior of "avoidance" equivalent to conscious level 31, for instance, is because the meaning of the relevant conscious level is "temporary feelings such as emotions."

Kitamura makes a hypothesis that the suppression of consciousness gives rise to the desire of and search for a behavior that is expected to bring about better results. This means that to remove the cause of the suppression of consciousness and expect better results, the consciousness of an individual with a better developed or more complicated physical body, from the viewpoint of phylogeny, is necessary.

I will explain how the robot then elevates its conscious level to a higher level and selects a behavior corresponding to that level.

When more than one behavior exists at the same conscious level, all candidate behaviors are evaluated (the procedure for this will be described later), and a behavior that is evaluated higher than the behavior immediately before, even though the difference may be marginal, is selected. It is possible that the robot steps down layers and selects a lower level. For example, a robot chasing an object at level 5 using a medium, upon finding a new object appearing abruptly in its vicinity, may lower its conscious level from 5 to 2 in an attempt to access this new object by selecting the "move" behavior.

It should be noted that when a robot is behaving at a certain conscious level, behaviors specified for all lower levels are included as subconsciousness (and this term is used here by the author in place of unconsciousness as used by Kitamura to distinguish between them). Moreover, when the conscious level is elevated, all previous conscious levels that were used are retained as subconsciousness. When stepping down through layers out of necessity, all consciousness existent at the upper levels is cleared and the consciousness of the relevant lower level emerges.

Next, I will explain the method for controlling the consciousness architecture proposed by Kitamura by citing his writings (Kitamura, 2000).

Assume that C_i is a function representing the strength of consciousness to change the conscious level. I_i is the function used to decide a behavior given conscious level i, where $i = 1, \ldots, 5$.

$$C_i = |\beta_i| + |E_i| + |\gamma_i| \qquad (8.1)$$

$$I_i = \beta_i + E_i + \gamma_i \qquad (8.2)$$

where β_i is perception, E_i expectation, and γ_i physiological perception parameter. Perception β_i takes a positive or a negative value when the object is "love" or "hate," respectively. The absolute value relates to the distance to the object. Expectation parameter E_i exists for conscious level 3 and above.

For $i = 3$, E_i is proportional to β_i for a given time period and is thereafter proportional to $-\beta_i$ if β_i does not increase. If perception β_i of an object is "love," expectation is set high for a certain time period and then set low thereafter. This means that if a good result is not obtained within the given time, expectation turns to disappointment.

For $i = 4$ or 5, E_i is proportional to the sum of present β_i and the previous β_i's considering the effect of oblivion. This means that the magnitude of historical obsession of the robot with various objects is taken into consideration.

Selection of behavior is described as follows. Behavior I_i selected by the robot at conscious level i satisfies the following equation:

$$I_i(t) > I_i(t-1) \qquad (8.3)$$

where t is current time.

When there is no behavior pattern satisfying Eq. 8.3 at conscious level i, the robot elevates its conscious level to $i + 1$. Parameters β_i and γ_i are used to characterize the robot. This includes, for example, emphasizing "love" or "hate" of objects and the state of starvation. It seems that these are currently being decided by the designers of respective robots.

Kitamura introduces four experiments in his book (Kitamura, 2000). One of them is explained here.

Robot H captures robot T in this experiment. Robot H has five conscious levels, whereas robot T has only three. From the viewpoint of phylogeny, robot H is physically superior to robot T, that is, its neural pathways are more sophisticated and functional. A relatively large obstacle and the nest of Robot T are prepared in the test area. Robot T is initially placed in the center of the test area and Robot H is set slightly away from Robot T. The positions of robots H and T, respectively, at time I are h_i and t_i.

The experiment starts with Robot T finding its favorite nest and approaches it by elevating its conscious level to 2 (position t_1). Robot H tracks robot T by elevating its conscious level to 4 (position h_1). Robot T takes a rest in its nest by lowering its conscious level. After resting a while, Robot T starts moving on receiving stimulus from its sensors and by elevating its conscious level to 2. Having lost sight of its target, and therefore having its current behavior suppressed, Robot H elevates its conscious level to 5 to start a series of behaviors of looking around for a medium and waiting for the target by hiding behind the medium (position h_2). Robot T discovers an obstacle and, because its behavior is thereby suppressed, elevates its conscious level to 3 to initiate an avoidance behavior. After having cleared the obstacle, Robot T lowers its conscious level to 2 to resume the move behavior. Through its sensor robot H catches sight of robot T coming out of the obstacle and chases and captures it (positions h_3 and t_3).

By observing this experiment, we know that Kitamura's conscious robot functions very well. Figure 8.4 shows the Locus of Ambush Behavior by Kitamura, which was partly modified by the author to describe this experiment (Kitamura, 2000).

8.2.1 *Summary and Observations*

Kitamura's achievements and unsolved problems are described as follows. He proposed a consciousness architecture based on Brooks' subsumption architecture. He hierarchized the levels of consciousness just like Brooks did with behaviors. In Kitamura's hierarchized consciousness, any conscious level subsumes all of its lower levels just like in Brooks' design where any behavior level subsumes all of its underlying behavior levels. The phylogenic complexity of living organisms is taken into consideration in Kitamura's hierarchized consciousness; simple organisms are assumed to have simple consciousness and are deployed on the lower levels, whereas

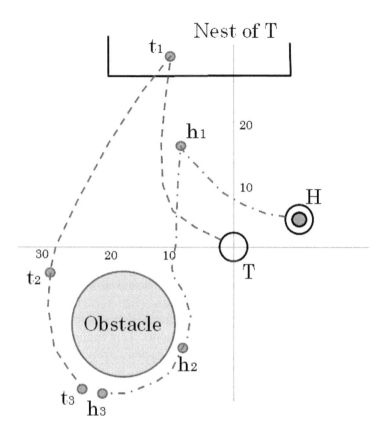

Figure 8.4. A rough sketch of Kitamura's experiments.

higher organisms are considered to have complex consciousness and are placed on the higher levels. Kitamura, thus, believes that all living organisms have a consciousness corresponding to their level of complexity.

Consciousness occurs, according to Kitamura, when transition to a higher conscious level occurs when the behavior at the current conscious level is suppressed. As many researchers agree, this is knowledge obtained from observing that when a human experiences that the current action is suddenly suppressed, the human's conscious activities are activated (i.e., they become conscious) and attempt to solve the problem.

Kitamura learned about conscious levels and their corresponding behaviors from philosopher Thao and applied it to the design of his consciousness architecture.

The author highly evaluates Kitamura's consciousness architecture, a machine consciousness theory, because it coincides well with the events obtained from phenomenological knowledge.

Brooks' standpoint is, however, basically behavioristic, and he does not acknowledge the conscious activities of humans and representations. It is therefore highly possible that Brooks would not accept Kitamura's layered consciousness and would disprove of it saying that everything can be explained by layered behaviors only.

Returning to our main theme, Kitamura describes the future course of development as follows. The first task to solve is related to the "emergence of I," that is, the problem of "first-person properties of consciousness." Kitamura believes that this problem can be solved by further increasing the levels of consciousness. According to Kitamura, the "emergence of I" occurs when consciousness has been elevated to the highest level where there is no further activity capable of enhancing the present comfort. This applies, Kitamura continues, to living organisms of low conscious levels as well.

Organisms at level 5, for example, can be conscious of the "emergence of I" but cannot speak of it simply because their consciousness is not high enough to engage in language activities.

Kitamura further says that it is difficult at this stage to correctly propose behaviors corresponding to conscious levels exceeding level 5, but such problems as symbolic processing can be solved by further elaborating conventional AI knowledge.

The duality of self-consciousness is described as follows. According to Kitamura, when "I" recall past experiences at the time when "I am emergent," the duality of consciousness that "I think of myself" may be said to be achieved. Regarding embodiment, Kitamura says it is a difficult technical task for a robot to become conscious that its limbs are its own. To solve this technical task, Kitamura continues, the robot needs to use the body schema of its own limbs. Kitamura says that the problems of the consciousness of others, feelings, and thought and the problem of chaos remain unsolved and will be continually studied in the future.

8.3 Jun Tani's Robot

Jun Tani built a conscious machine based on the belief that the conflict between the external and internal worlds gives rise to self-consciousness. He developed this idea after studying the theories of Martin Heidegger and many other philosophers and psychologists.

The external world means the environment of the robot as captured by its sensors. The internal world consists of a series of artificial neural networks accommodated inside the machine (Fig. 8.5).

Tani proposed an open dynamic structure featuring multiplexed neural networks and built it into his conscious machine (Fig. 8.6). The circuitry consists of a group of neural units (a) playing the role of a window for interaction between external and internal worlds and a cognition and prediction unit (b) including context loops. The context loops are used to predict the next occurring movements and objects based on past experience and have a memory to remember events in the external world.

Section b comprises a set of recurrent neural networks (RNN) responsible for two processes: cognizing sensor inputs and predicting the next occurring sensor values.

Let us assume that Tani's machine has learned enough in the external world, and as a result, it is expected to perform smooth movements. Actually, the machine has succeeded in learning smooth movements. Section b of the machine constantly outputs prediction values of "What and Where" to section a for collation with the present values. Smooth movement is assured when the prediction and external world coincide with each other. If the prediction fails to match, the robot's movement is not smooth. Look at the internal state of the robot at this time. You will see sections a and b are alternately stable and unstable. It is stable when the prediction agrees with the external world and unstable when it fails to agree.

As discussed in Chapter 7, the recurrent neural network has four states: divergence, convergence, orbiting, and chaos. We can safely exclude divergence because this robot has sufficiently learned in the external environment. The aforementioned stable state is equal to convergence. If RNNs repeat convergence even though their state is incessantly variable as the machine moves, the smooth

External world

Figure 8.5. External and internal worlds.

movement of the machine is achieved. Orbiting, including quasi-orbiting, can be said to be a quasi-stable state. After all, the machine can continue to move as long as stable and quasi-stable conditions are maintained in the real world. Continuous stable movement thus

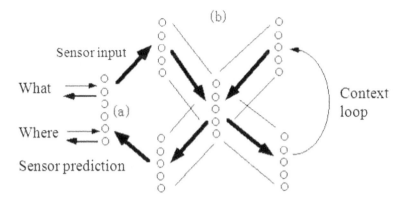

Figure 8.6. Open dynamic architecture.

means agreement between the predictions and the external world. The RNNs, while learning information about the external world that they successfully predicted, further strengthen their capacity to predict.

What happens, then, if we arbitrarily change the external world which the robot has learned? The RNNs get unstable and enter a state called chaos. When the machine enters the chaos state, the orbit is seriously distorted. In short, prediction and reality do not match, and it is no longer possible to predict. If this state is allowed to continue, the RNNs cannot escape from chaos. Tani at this point incorporates an arbitrator mechanism in his architecture. The arbitrator reduces the impact of the predicted values when in chaos. As a result, the machine is capable of re-learning information about the external world. According to Tani's interpretation, self-consciousness occurs in the machine when it is in chaos. Tani also explains that the state where the machine runs stably corresponds to non self-consciousness. He describes self-consciousness referring to Heidegger's hammer.

When a carpenter uses his hammer without any difficulty, the carpenter and the hammer are in unity, but when problems recur, they are separated and the carpenter is conscious of the hammer and he himself becomes self-conscious. This theory is supported by many phenomenological philosophers and psychologists.

8.3.1 *Summary and Observations*

Tani's conscious robot consists of multiplexed recurrent neural networks. He designed the open dynamic structure. In its basic design concept, Tani's robot captures information about the external environment to engage in cognition activities, while at the same time information about the external environment is predicted inside the machine. The robot moves in the environment smoothly when the result of the current cognition coincides with the internal prediction. If the prediction is different, the internal state of the robot enters chaos, and the robot moves awkwardly.

Tani interpreted that from a phenomenological standpoint, self-consciousness emerges when the robot enters the chaotic state. He relates the chaotic state in the open dynamic structure to the emergence of self-consciousness. However, is it possible to explain all the functions of human consciousness by relating the chaotic state to self-consciousness? There remains a serious question. For example, how would Tani explain the function of consciousness for willfully exploring objects based on chaos? If self-consciousness arises from the chaotic state as he argues, the above example must also be explained from the aspect of chaos.

8.4 Mitsuo Kawato's Examples

Mitsuo Kawato is a leading researcher of computational theory of the brain in Japan. His central theme of study is the elucidation of the human brain by a computational approach.

His famous research papers include, *Motor Learning Model of the Cerebellum*, *Computational Theory of Motor Control*, and *Computational Theory of Vision*. Of the wide range of his studies, his research related to consciousness is introduced in this section.

In his study of the brain, Kawato challenges some long-standing problems. One of them is called the binding problem. The author personally believes that this problem caused Kawato to decide to study consciousness. The binding problem originally emerged from the findings of the physiological study of the brain. It refers to a mysterious phenomenon related to the coincidence of the physiological fact that for an object that a human cognizes, the brain analyzes the object at the cerebral cortices in detail using the nervous system with the metaphysical fact that the human cognizes the object at the same time as the above physiological process goes on.

Assume, for example, you hold an apple in one of your hands and at the same time you look at the apple (Fig. 8.7). The brain sends commands to your arm and hand to drive muscles to keep holding the apple (Fig. 8.7a), while, at the same time, the brain receives information on the weight of the apple, its texture, and the somatic sensation of the driven muscles via the somatosensory system. At the same time, the eyes (Fig. 8.7b) capture the image of the apple and send the information from the retinas to the brain via the optic nerve system. The image of the apple is then transmitted to the visual areas V1, V2, V3, V4, etc. (see Chapter 3) of the brain and further to the modules that cognize form and color, and movement of the apple, respectively. The modules exchange information with the cerebral cortex and conclude the analysis of the apple (Fig. 8.7). It is obvious that information is exchanged with the memory modules. To hold the apple, various other areas of the brain related to control of the hands and arms are likewise assumed to be at work.

Given the physiological knowledge that the cognition of an object is achieved when visual information is transmitted to the brain through the optic nerve system and via various modules, the following question arises: When does cognition or awareness of holding an apple in the hand occur.

It should be noted here that "the brain cognizes" is different from "realization." Personally I believe that "realization" refers to "becoming aware of" or "something emerges on the consciousness." Even when you cognize "holding an apple," the cognition can be subconscious.

Let us return to the binding problem. Referring to the above example, the information on the apple as it is seen by a human goes to various modules via V1 through V4 until it reaches the columns of the cerebral cortex, where the analysis of the apple concludes. Since information about the apple is functionally divided into pieces, there arises the question how these small pieces of information are analyzed and finally assembled, or "bound," to form a cognition.

Where and when does this integration occur?

Figure 8.7. Binding problem and bidirectionality theory. Cited from Kawato's drawing in reference Kawato (1996), partly modified by the author to facilitate the explanation.

The result of the analysis of these segmented pieces of information may be used to explain that simple combinations lead to cognition as a kind of solution. It is obvious, however, that the number of possible combinations can be astronomical, and thus this explanation has no possibility of being accepted as a solution.

In 1988, Christoph von der Malsburg announced his theory of synchronous firing to solve the binding problem. According to this theory, the identity of an object is shown when groups of segmented information belonging to the object are synchronously fired.

Kawato posed three questions regarding this theory.

First, to identify more than one object by synchronous firing, the accuracy (i.e., window) of synchronization must be about 1 ms or 2 to 3 ms at the largest (Kawato, 1996), and he denied the possibility of the brain achieving this.

Second, a combinatorial explosion is not improved by this theory.

Last, there is no answer to the question, "Who will witness the synchronization?" Something like a central control center would be required to organize a multitude of modules, and this would bring us back to the first question.

Kawato concludes, "After all, any of the three proposed theoretical possibilities (a single monster integration center, module binding, and synchronous firing) fails, at least at the present time, to describe how a unified image is successfully formed and solve the binding problem, given the large number of visual areas and sparse binding (parallel flow of information in a hierarchical structure)" (Kawato, 1996).

Kawato proposes the bidirectionality theory that is allegedly possible to solve, even though not now, all of these problems. The bidirectionality theory has been developed to explain the high-speed processing of the human visual function. This theory is explained referring to the already-discussed example of an apple. When the visual information is segmented and transmitted to the cerebral cortex, the same information always flows in the reverse direction, and the sameness of the forward and backward information is maintained (dashed lines in Fig. 8.7). Neural pathways from a lower to a higher area are analytical paths (solid lines in Fig. 8.7). Analysis is generally a kind of modeling; so Kawato calls this operation forward modeling. The information from the result of the analysis is then fed back to the lower area via a neural pathway extending in the opposite direction from the higher level (dashed lines). The role of this reverse neural pathway is to integrate the result of the analysis and create an approximate model called a reverse model in the lower area, according to Kawato.

To provide the theoretical grounds, Kawato refers to the bidirectional binding between areas of the cerebral cortex, which is an established anatomical fact about the brain. He argues that his theory solves the first and the second of the three problems already mentioned. Specifically, the integration of information can be finally performed in relatively low areas according to his theory. More detailed information, when necessary, is obtainable at any time by accessing the higher level cortices. This way, the second problem of the combinatorial explosion can be avoided. The binding of all areas is necessary if you follow the theory of synchronous firing. It is clear that this problem is also avoided in the bidirectionality theory. Kawato does not seem to discuss the third problem.

I will now describe how consciousness is addressed in the bidirectionality theory. Kawato applies his bidirectionality theory to the problem of not only visual processing but also sensorimotor integration. His study is considered a sort of imitation learning. This means that he tries to analyze information obtained by the vision system (forward model) and integrate the results in the motor system (reverse model). The idea of bidirectionality can be used because "one moves as one sees" is the essence of imitation learning. Kawato then extends the concept of forward and reverse models (Fig. 8.8). I have partly modified his drawing here to facilitate understanding.

Section A in Fig. 8.8 represents a set of sensors for hearing, vision, and the like. Section B is a set of drive systems for the limbs. In the middle part of the figure, there are neural networks and modules M_i conceived in the bidirectionality theory. Modules M1, M2, M3, ..., for example, are responsible for analysis and integration of perception, while modules M4, M5, M6, ... are related to analysis and integration of motion.

Kawato does not describe the connection between the perception and the motor system. However, if sensorimotor integration is achieved by the bidirectionality theory as asserted by him, Kawato's argument that the bidirectional forward and reverse models can generalize other individuals, one's own brain, and part of it, and society existing in the external world is indeed persuasive enough. Kawato also asserts that his bidirectionality theory has the merit of cognizing or recognizing the patterns that are created by motor control. He also discusses the problems of consciousness. According to him, "Consciousness exists to accelerate integration of the modules in the brain." He says, "Self-consciousness may be partly explained as 'a certain part of the brain constructs the forward or reverse models of interactions for a number of modules" (Kawato,

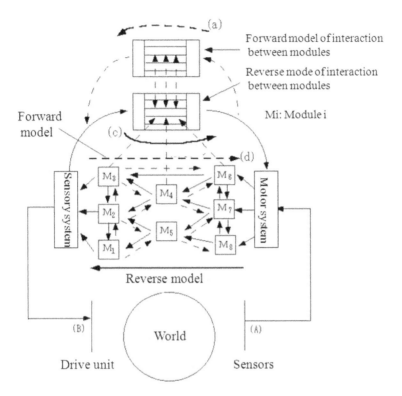

Figure 8.8. Bidirectionality theory and selfconsciousness (Kawato's drawing was simplified by the author for this description).

1996). Furthermore, he says, "The role of the forward and reverse models of interaction between the modules is to set favorable initial conditions for the relaxation algorithm by an intermediary of a number of modules, and accelerate the relaxation algorithm for the integration of the modules." He explains, "When our consciousness works normally, we feel the world's only perception image and are executing a consistent action plan."

It is not difficult, according to Kawato, to model interactions between modules. The problem is if the total memory of the interaction models becomes equal to or exceeds that of the main bidirectional modules, the merits of convergence and high-speed processing are lost, and hence the interaction models must be poor. "Forward models of interactions between modules can be regarded as the approximate models of other areas of the brain," Kawato says, "Based on this hypothesis, we can say that consciousness is a process of approximating the huge volume of parallel calculations that occur subconsciously to resolve the ill-posedness of sensorimotor integration with extremely simplified false serial calculations." He further argues, "Forward models of module integration, generally, are similar to reverse models of the external world because they convert sensory inputs into motor outputs. Likewise, reverse models of module integration can be said to be forward models of the external world. These computations circulating through the forward and reverse models make it possible to emulate one's own thoughts, predict one's own behavior and perform the act of introspection." Emulation here means to run part or the whole of a system on other software or hardware that behaves like the original system.

Kawato's writing is not so easy to follow. I will interpret it as I understand it, risking the claim of misunderstanding. Forward models of interaction (Fig. 8.8a) are approximations of reverse models (Fig. 8.8b) of sensorimotor bidirectionality modules, and the reverse models (Fig. 8.6c) of interaction are approximations of forward models (Fig. 8.8d) of the bidirectionality modules. These approximate forward and reverse models are considered to be situated above the real-world models (forward and reverse models of the sensorimotor bidirectionality modules) and behave as an emulator that approximates the real world as a two-tier world. Kawato's argument that computations circulating through the approximate forward and reverse models, which are sparse due to the two-tier structure, enable the "emulation of one's own thought," "prediction of one's own behavior," and "introspection" is understandable as an idea but lacks concreteness.

Kawato further discusses the flow of consciousness.

Each forward or reverse model of module interactions comprises three parts: interface to the sensory module presumably located in the prefrontal area, interface to the motor area assumed to be partly located in the basal ganglion, and the model itself which is said to reside on the outside of the cerebellar hemisphere, and is divided into several micro-zones as shown with the vertical lines in the figure. This means that consciousness has just one stream but the direction of the stream of consciousness is controllable. When driving a car, for example, you can converse with passengers, listen to music or concentrate on driving. This is because although the interface to the sensory and motor modules is just one route, several models are prepared to cope with certain possible combinations of modules (Kawato, 1996).

Kawato's description lacks clarity, but he seems to mean that the flow of consciousness (the reason why "a driver is capable of conversing with passengers") is related to the micro-zones located in the interaction modules.

8.4.1 *Summary and Observations*

Kawato studies the integration of the human sensory and motor systems. He needs to solve the long-standing "binding problem" that has been annoying brain researchers. To solve this problem, Kawato proposes the bidirectionality model for sensorimotor integration. According to Kawato, the proposed method is capable of suppressing a combinatorial explosion and other problems compared with conventional solutions. He further proposes the idea of module interaction models as a higher level of a bidirectionality model to accelerate the processing speed of various modules in the bidirectionality model. His interaction models are approximations of the already-mentioned bidirectionality models. He explains that the interaction models can approximately reproduce the functions of the bidirectionality models He explains that the "emulation of one's own thought," "prediction of one's own behavior," and "introspection" are enabled because the interaction models approximately reproduce the bidirectionality models. He further says that self-consciousness may be explained in like manner.

According to Kawato, consciousness is a process of approximating the huge volume of parallel calculations of sensorimotor integration that occur subconsciously with simplified false serial calculations. He says that the flow of consciousness is related to the microzones of the interaction models.

The author acknowledges the value of Kawato's bold hypothesis introducing bidirectionality models to explain consciousness but at the same time cannot see any concrete explanations. Kawato seems to have not yet formed a consistent theory about the function of human consciousness.

8.5 Cynthia Lynn Breazeal's Kismet

Cynthia Lynn Breazeal developed, in cooperation with Brooks, a robot named Kismet, which is capable of interacting with humans in real time. The humanoid head robot can move its eyes, eyebrows, mouth, and neck (Fig. 8.9), and it listens and talks. This sociable robot understands feelings using information from differences in height, pitch, etc., and expresses its feelings vocally. Its facial expressions also show feelings. The drive mechanism makes use of the psychological knowledge of the Theory of Mind. An outline of the circuit developed on the basis of this theory is shown in Fig. 8.10.

With these circuits, the robot reacts by driving motors in response to the stimuli given to the system via sensor inputs. Various

Figure 8.9. Head robot Kismet.

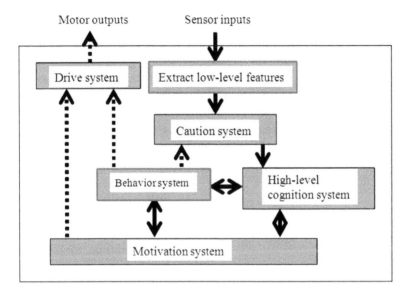

Figure 8.10. Outline of Kismet's drive system.

subsystems are layered and implement their respective functions. As the essential part of the system, the information obtained by the sensors is subject to the extraction of low-level features. On the basis of the resultant information, the high-level cognition system is activated by the Caution system, that is, the cognition process is a two-stage process. There are two closed loops for processing information internally. In one circuit, information circulates through the Caution, High-Level Cognition, and Behavior systems. In the other circuit, information runs through the Motivation, Behavior, and High-Level Cognition systems. In the former circuit, if any information has a feature to note, the robot turns to the object or the vision cameras are rotated by drive motors while using the Behavior system as if a human were looking at an object and watching it intently. This is intended to achieve a higher-level cognition. In the latter circuit, the robot's attention is given to already-learned information such as color and shape, to achieve high-level cognition. The two circuits each have their own roles. The Motivation system is the highest-level layer in this system.

8.5.1 *Summary and Observations*

What is most interesting about this system is the parallel use of bottom-up processing for driving the robot based on information transmitted from sensors and top-town processing using information from the Motivation system, which is located at the top of the subsystem hierarchy. The robot achieves interaction with humans smoothly while solving the problem of competition of the two different types of processing.

What interests me is to find out where in these circuits lies the essence of consciousness.

Chapter 9

New Architecture of Robot Consciousness and the Robot Mind

This chapter introduces the conscious robot that passed our mirror image cognition tests for the first time in the world (Fig. 9.1). Mirror image cognition means to "cognize the image of the self in a mirror." It is considered to be the source of self-consciousness.

What is the mechanism of human consciousness?

This chapter discusses this mechanism.

The electronic edition of the Discovery Channel in the United States reported on this robot on December 21, 2005. Within two or three weeks after the report, about 26,000 secondary articles and references to this robot appeared on Web sites all over the world. The robot was discussed on many Web forums, and the information spread throughout the Internet, where many comments of both praise and refutation were seen.

The effect of the publicity on the Discovery Channel was enormous. Four months after its appearance, the conscious robot was listed in the second place in Google search results for the keyword "self aware," which logged about 132 million hits. The conscious robot was, of course, at the top of the list related to scientific research.

Creation of a Conscious Robot: Mirror Image Cognition and Self-Awareness
Junichi Takeno
Copyright © 2013 Pan Stanford Publishing Pte. Ltd.
ISBN 978-981-4364-49-2 (Hardcover), 978-981-4364-50-8 (eBook)
www.panstanford.com

Figure 9.1. Conscious robot that passed the mirror image cognition tests.

9.1 Introduction

What neural circuits are used, and how are they used, to make up the mechanism of human reason? How is the mechanism of emotions and feelings related to reason? What is consciousness and self-consciousness? What is the difference between subconsciousness and explicit consciousness? What is the meaning of "to feel?" What is free will?

The author aims to answer these questions. I would like to innovate conventional artificial intelligence (AI), which is presently at a deadlock. I would also like to construct a totally new AI, which I call artificial consciousness (AC). I am also trying to construct a model of the human brain. I do not mean to use the risky anatomical analysis of the brain, but I will construct a mechanism to represent human consciousness in an embodied robot. The rich knowledge I expect to derive from my research will help resolve the mysteries of the brain. If we could uncover these mysteries, we could reap immeasurable benefits from our efforts.

First, AC-based machines will be designed and developed to ever higher levels. Conventional automobiles, for example, have been designed to follow the driver's instructions, whereas an AC-based automobile will function as though a dedicated operator were driving it in place of the human driver. The car runs automatically, always checking its safety, and observing the instructions given by the driver from time to time. Even if the driver were to faint for some reason, the car would continue running safely until it slows down and stops at a nearby safe place, or may continue on and arrive at the destination if time is not a critical factor. The difference from conventional vehicles is the intrinsic safety derived from the autonomy of the car and enhanced comfort. The driver's stress would be greatly reduced.

Second, AC technology will create a simultaneous interpreter robot that speaks natural languages. The robot interpreter will engage in simultaneous interpretation, the quality of which would surpass human interpreters. It understands homonyms from the context and appreciates and returns with a joke and irony when appropriate. In a hospital, the AC robot selects and avoids speaking upsetting words and phrases that may touch on diseases.

Third, the AC robot will attend to old people in need of care and physically handicapped persons and provide physical and mental care for elderly living alone. The robot will take care of these people while being conscious of their mental and physical conditions. It reaches out and supports them to walk safely. The AC robots will support people in wheelchairs and bedridden patients by speaking with them gently, encouraging, listening and then advising, or otherwise providing mental support.

Fourth, the AC robot will offer ideas for survival in the event that a human is in trouble and continue to support the person as long as its energy lasts.

Fifth, it is possible to develop artificial limbs that human users will feel are a part of their own body.

Sixth, the AC robot has the function of self-consciousness in principle and always monitors its own activities. When any problem occurs in the AC system, the relevant information is used to anticipate and prevent catastrophic breakdown of the robot's overall system to minimize damage. There is also the possibility of self-repair. With all these features, AC robots will contribute to the safety of people all over the world.

Lastly, it would be possible to discover a medical treatment for people suffering from serious mental illnesses such as schizophrenia. If the mechanism of consciousness can be eventually elucidated, a technology to resuscitate lost consciousness might be developed.

The author's idea is that "consistency of cognition and behavior is the origin of human consciousness." Cognition means to understand. The idea, therefore, means that the function of consciousness is to behave and to understand the behavior to maintain consistent information.

The author has developed, on the basis this idea, a module to build consciousness on a robot, which is called Module of Nerves for Advanced Dynamics (MoNAD). The consciousness system is constructed by arranging multiple MoNADs in a layered fashion.

9.1.1 *Research on Consciousness and Cognitism*

The first thinker to be introduced when it comes to human consciousness is French philosopher Rene Descartes (1596–1650), who was briefly mentioned earlier. He was the first person to assert the existence of the self on the grounds that "I feel myself" by saying, "I think, therefore I am" (Descartes, 1997). His approach is called dualism, which asserts that the mind and the body are separate entities.

Later, German mathematician and philosopher Gottfried Leibniz (1646–1716) attempted to treat the body and the spirit integrally in his Monadology (Leibnitz, 1714).

No scholar or researcher has ever reached a unified view of the body and mind based on physical understanding.

When a human is conscious of something, he or she feels it by himself or herself, and nobody else is involved. This is because to be conscious is a subjective matter and, some may argue, rejects analysis by science. To study consciousness through a scientific approach, a technique was devised to describe the reactions of humans to stimuli. The belief in this technique is called behaviorism.

Since about 1960, scientific findings have been announced one after another showing that complex processing is going on in the human brain. This made it necessary to explain brain activities and stimulated researchers to this end. One of these scientific findings reads like this: Assume, for example, a mathematical question of whether certain numbers are included or not in a given set of numbers. It was revealed that the time required to answer the question became longer as the digits of the numbers increase (Sternberg, 1966). The theory that accepts the existence of complex processing in the brain is called cognitivism.

Later, many advanced scientific equipment and systems to measure brain activities of humans were developed, including electroencephalography (EEG) and functional magnetic resonance imaging (fMRI). These technical advances assisted in the discovery of more scientific grounds for the existence of complex processing in the brain.

As of today, it is known that

(1) The brain engages in complex processing when solving a complex task.
(2) Specific areas of the brain are activated.

Behaviorism has evolved into cognitivism as a result of the development of science. Specifically, the description of human reactions to external stimuli has been gradually replaced by research on the internal processing of the brain. There still exists, however, criticism that subjective problems based on the sense of the self such as consciousness and the mind still lack scientific grounds.

9.1.2 *Husserl's Phenomenology*

Phenomenology proposed by Edmund Husserl (German philosopher, 1859–1938), mentioned earlier, is introduced in this section again (Husserl, *Die Idee der Phaenomenologie*). Phenomenology is a difficult philosophical idea but is described here as the author understands it.

Phenomenology is a philosophical method to establish a thought that has not been established. Simply put, a hypothesis is first proposed regarding an unknown world and studied by comparing it with phenomena emerging in the real world. A new hypothesis, agreeable to a larger number of people in the field, is presented on the basis of the results of the study. The hypothesis may at first be a subjective idea of a researcher, but by repeating the above process, an agreement between subjectivity and objectivity is eventually reached. At this time, the phenomena have acquired scientific grounds. By using this philosophical method called phenomenology as advocated by Husserl, one can thus establish scientific grounds for as yet unproved phenomena by following scientific processes. This method, also known as scientific positivism, is actually a representative method used by researchers in the study of unknown worlds.

Consciousness and the mind are subjects of study for which no scientific grounds have been established until now, and that is why the phenomenological method is actively being used to study them.

9.1.3 Definition of Consciousness

The author selected consciousness as the theme of his study, believing that the mind is the embodiment of consciousness. The author further defines consciousness as a function to perform cognition and behavior consistently, the details of which will be explained later.

What does it mean to become aware?

As a result of phenomenological observation, the author believes that to become aware means to realize that one is doing something.

To feel is a phenomenon that occurs when one becomes aware. The author believes that to feel and to become aware are different because to say that one becomes aware of feeling is understandable, whereas to say that one feels becoming aware is a somewhat uncomfortable usage of language. This suggests that in the human brain, the concept of feeling is at a lower level than that of becoming aware, and the two are independent. Thinking in this way, it seems that to become aware is synonymous with to be conscious.

What, then, is consciousness?

The author uses Husserl's 10 features of the functions of human consciousness consciousness (Husserl, *The Essential Husserl: Basic Writings in Transcendental Phenomenology*), for which scientific grounds have been established by Husserl himself using the phenomenology he proposed. The 10 features have been demonstrated by Prof. Tadashi Kitamura at the Kyushu Institute of Technology (Kitamura, 2000; Kitamura *et al.*, 2000).

9.1.4 Features of Consciousness

The 10 features of consciousness are described as follows:

(1) First-person property: First-person property is the sense that one is performing all things, i.e., a belief in the existence of the self or mind–body monism.

(2) Orientation: Orientation means that consciousness is always directed toward something.

(3) Relationship between action and result; duality of self-consciousness: Relationship between action and result refers to the chaining of an action and its result. Humans always feel a relationship that a certain action has a specific result. The duality of self-consciousness means, in other words, that one is aware of oneself.

(4) Expectation: Expectation indicates that humans are always predicting the immediate future.

(5) Function of determination and conviction: The function of determination and conviction is identical with the belief in the existence of things.

(6) Embodiment: Embodiment is the feature that the body is part of the self. All of us are conscious that our body is part of our self. The importance of embodiment in consciousness is addressed in detail in the *Phenomenology of Perception*, by Maurice Merleau-Ponty (1908–1961) (Merleau-Ponty, 1945).

(7) Consciousness of others: Consciousness of others is the feature that enables us to discriminate our self from others.

(8) Emotional thought: Emotional thought is the feature that reason is related to emotion and feelings.

(9) Chaos: Chaos means that consciousness is ceaselessly out of balance.

(10) Emotion: Emotion refers to qualia of consciousness. The human senses of taste, hearing, smell, color, pain, etc., are deeply related to qualia.

A consciousness system, when proposed, should be tested using these 10 features proposed by Husserl. How far the proposed system can describe these features is an important index in the evaluation of the proposed system.

The consciousness system proposed by the author can describe all of these 10 features. Only one item, No. 10, emotion, has not been achieved on the robot constructed by the author.

Figure 9.2. Imitation of Facial Expressions by Infant (images created by the author).

9.1.5 *Important Research Examples Related to Consciousness*

Let me first introduce an example shown in the research of imitation behavior by infants as reported by Meltzoff and Moore (1977). They discovered that a four-week-old infant imitated facial expressions presented in front of the infant, such as sticking out the tongue, opening the mouth, and pursing the lips. According to them, infants are capable of imitation as soon as they are born. An imitation scheme is already wired in the neurons of the brain when a baby is born, i.e., imitation is an innate ability (Fig. 9.2).

There is a case study in imitation behavior (Lhermitte *et al.*, 1986) which was observed on a patient with partial damage to the frontal lobe (Fig. 9.3). The patient in question was unable to suppress imitating the behavior of the person sitting in front of him. When the person in front of the patient reached out for a pair of eyeglasses and put them on, the patient immediately behaved in the same way. This case underpins the hypothesis that imitation is one of the fundamental behaviors of humans.

Mimesis theory is worthy of mentioning here as another discussion of imitation. According to the theory, imitation was an important means to bring people together in an age when humans had not yet developed languages (Donald, 1991). This hypothesis suggests that imitation was an important function of humans in the history of human evolution.

There is another important research result. It is the great discovery of mirror neurons (Gallese *et al.*, 1996). by Prof. Giacomo

Figure 9.3.   Examples of imitation behavior (from Lhermitte *et al.* (1986)).

Rizzolatti, a neuroscientist at the University of Parma (Arbib, 2002; Rizzolatti and Sinigaglia, 2006). The discovery of mirror neurons showed the existence of common neurons related to cognition and behavior in the brains of monkeys and humans.

These numerous findings show that the mechanism of imitation is critically important for human consciousness. This leads to the fact that when one claims that the consciousness system they propose is capable of simulating human consciousness, then the proposed system must possess the function of imitation as a basic feature.

9.2 Proposed Concept Model of Consciousness

The computation model of the consciousness module proposed by the author is introduced in this chapter. The model is described in a format allowing creation on a computer. The correlation with Husserl's 10 features is also explained. The reason why the self and others can be discriminated and why imitation is possible is also discussed.

9.2.1 *Artificial Consciousness and Design of the Mind*

Consciousness has been an important research theme in recent years and is actively being studied in many disciplines, including philosophy, psychology, brain science, cognitive psychology, and robotics.

Consciousness is defined variously as the working of the mind that recognizes and aspires, the state of mind that one knows what one is doing now, and something that includes and underlies all workings of our knowledge, feelings, and will. The mind is used in a broader sense than consciousness and is defined as the whole of knowledge, feelings, and will, whereas consciousness is generally understood to be something underlying the mind.

As mentioned earlier, the study of mind started with the philosophical discussions of Descartes, who launched mind–body dualism. The study of the mind evolved with research on the subconscious by Freud and other psychologists.

To facilitate scientific discussion of the subjective matter of the mind, phenomenologists, including Husserl and Merleau-Ponty, offered the necessary methodologies.

Brain scientists are engaged in researching the scheme of the consciousness and the mind, with the brain as the central theme, by directly stimulating the brain or by monitoring activities of the brain with the help of PET, MRI, and other investigation systems. Despite these numerous efforts, the location of consciousness and the mind has not been identified.

The discovery of mirror neurons is considered very important among all these achievements by scientists. The existence of mirror neurons is important because their existence supports the hypothesis that higher organisms learn by imitation.

The connectionists, including Prof. Daniel C. Dennett at Tufts University, argue that consciousness can be created artificially (Dennett, 1991) although concrete plans have not yet been shown.

Brooks and his colleagues at MIT have created COG, Kismet, and other robots to study natural learning systems through the interaction between infants and the environment and expect that consciousness and the mind will emerge from these learning procedures.

While partly agreeing with the ideas of these evolutionary robot researchers, the author believes that rather than attempting to create consciousness with evolutional techniques, we should develop artificial cell networks step by step while defining a new paradigm regarding consciousness. As mentioned before, the author doubts that consciousness can be created using evolutional techniques.

I believe that a new paradigm must be established to generate consciousness in a machine. Descartes advocated a mind–body dualism, but my standpoint is a materialistic mind–body monism. I believe that something like human consciousness can be constructed in a machine. A robot with something like consciousness would behave as though it had a will.

9.2.2 *Expectations for a New Paradigm*

The author believes that the imitation function and the function to distinguish between the self and others are important factors of human consciousness. Humans can distinguish the self from others using a function that discriminates between the self and others. They learn the behavior of others by just looking at them, thanks to the imitation function. These two functions are the prime mover in the evolution of humans into higher organisms.

The research studies by Meltzoff *et al.* have shown that infants instantly imitate the behavior of their mothers. It was quite strange for the researchers to learn that the infants performed the same behavior in a short period of time (Meltzoff and Moore, 1977).

When a mother sticks out her tongue, the child instantly does the same. The child seems to move to action immediately by just seeing the mother. When I touch the keyboard of a computer, I mostly strike the keys subconsciously but I never think that the fingers striking the keys are those of someone other than myself. When a person resembling me stands nearby, I never mistake him for myself.

Self-consciousness that tells that the image of oneself reflected in a mirror is the self seems to be deeply related to the function of distinguishing between the self and others.

The author is trying to discover the source of consciousness.

There are many researchers studying the problem of "What is consciousness?" Nevertheless, no assertion has ever been acknowledged to be decisive enough to solve this problem. This is because a unified paradigm regarding the phenomena of human consciousness had been missing.

The author has devised a new paradigm of consistency of cognition and behavior as the source of human consciousness.

This chapter shows that the new paradigm can describe almost all of the phenomena of human consciousness in a computationally feasible manner. The author's arguments may contain, in various points, findings that have already been discussed by other researchers, but what is asserted by the author as a whole is a unified paradigm capable of explaining many phenomena related to human consciousness and the mind.

The underlying idea of the system introduced herein is the construction of neural networks essentially defined between inputs and outputs based on this new paradigm. The system continually evolves by capturing information about the surrounding environment.

The author believes that thought is generated from conscious activities, i.e., a stream of consciousness. Consciousness here includes the subconscious. Explicit consciousness, generally referred as simply consciousness, occurs when the subconscious satisfies certain conditions, which will be discussed in detail later.

The mind is activated by the stream of consciousness and the accompanying stream of *kansei*. *Kansei* is a Japanese word that means both emotion and feelings. Consciousness is considered to occur within the brain according to the internal state of the brain and on receiving stimuli from the body and external environment. The new paradigm is capable of explaining the root source of consciousness and also describes imitation learning, distinction between the self and others, self-consciousness, and qualia. It can also describe imaging and creativeness. Furthermore, this new paradigm makes it possible to draw a distinction between sense and hallucination.

The methodology based on the author's paradigm resembles Minsky's idea that in the human brain, higher-level areas have evolved biologically from the old areas, i.e., the old areas are

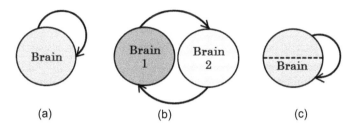

Figure 9.4. Development of the brain and self-recurrence.

supervised by the newly evolved areas. This idea may remind the reader of a centrally controlled system where the new brain dictates the activity of the old brain, but the author's methodology is different. The author's methodology is the same to the extent that the new and old brain areas co-exist, but the difference is that rather than the new brain controlling the old brain, information circulates between them. This idea may not sound smart, but using this idea it is easy to explain the function of self-consciousness that is defined as the state of being conscious of one's own consciousness.

It is easier to understand the model of self-consciousness by looking at Fig. 9.4a and then Fig. 9.4b. In Fig. 9.4a, the self is looking at the self, i.e., the self-loop type. This is a typical self-recurrent model. In Fig. 9.4b, the self is looking at others, i.e., the co-routine type.

The reason why the author selects the co-routine type is the difficulty of explaining the internal circulation of information using Fig. 9.4a. With the co-routine type, it is easier to understand that information circulates through both areas. If this description is possible and successful, we may move to the smarter self-loop type.

When one of the two routines is combined with the other (Fig. 9.4c), the result is the same as the model shown in Fig. 9.4a. Accordingly, the author hereby declares that he is not using the centrally controlled model.

9.2.3 *Where Does Consciousness Come From?*

What kind of state is it in which we are conscious of something?

I am now writing letters on a notebook with a pen. Or I am thinking of my family, solving math problems, walking down a street in Paris, etc. It is appropriate to express this awareness as being conscious of something. It is also possible that I was walking while thinking about my family, but I was not conscious of walking. When this phenomenon is to be written into a computer program by analogy, a number of software agents are working in the computer and one of them is selected by means of some function.

Even if the computer displays a message on its screen that one of the simultaneously running multi-tasking programs is currently active, it would be difficult to say that the computer now possesses a certain consciousness. To overcome this difficulty, some researchers argue that consciousness requires embodiment (Pfeifer and Sceider, 2001) but this proposition would be promptly refuted by the opinion that computers can also have embodiment such as by the keyboard and the hard disk. A computer may be aware, through its functions, that its hard disk is now being accessed. Even so, it is difficult to accept that consciousness occurred in the computer. Why?

The author believes it is possible to define the relationship between the operation of the computer's hard disk and consciousness, but this, by no means, indicates that I am ready to admit the occurrence of consciousness in the computer. This is because the existence of consciousness is a subjective phenomenon felt only by an individual or by myself. I just feel that others also have consciousness as I do, but its existence is difficult to prove.

The mirror neurons described earlier seem to make us believe that other people also have consciousness. Since computers do not have a function similar to mirror neurons, it is natural that we cannot feel that computers have consciousness.

Some researchers thought about relationship between Turing tests and the human consciousness (Mogi and Taya, 2003). This means that they expect the action of the mirror neurons of humans. From these observations, one might say that for humans to be able to readily feel the existence of consciousness in a machine, the machine must constantly stimulate the mirror neurons of humans. This means that to create consciousness similar to that of humans in machines, the functions of mirror neurons must be built into the machine.

A clear definition of consciousness capable of describing the various functions of human consciousness must be established on the basis of the functions of mirror neurons. Since mirror neurons are related to the smooth perception of consciousness in others, the shape of the machine may have to be a humanoid or human-like robot. In the case of machines that are not human in shape, the existence of consciousness may be judged by whether the machine physically has the functions described in the above definition of consciousness.

9.2.4 *How Do We Define Consciousness?*

Let us consider the existence of consciousness referring to an example: The sound of a voice is input into the ears, and in response to that stimuli, the action of speech from the mouth is performed.

This example does not spoil generality. The area responsible for consciousness is assumed to be located in the region with the question mark in Fig. 9.5a. This is the area where neural networks exist.

The above thinking is behavioristic because the description reads that inputs entail outputs. If we apply an evolutionary algorithm to this region, neural networks with the functions of consciousness can be created in this region at some time.

Given the complexity of the phenomena of consciousness, it is difficult in reality to use the evolutionary approach and determine what functions are appropriate for creating neural networks equivalent to consciousness. If consciousness consists of a redundant system as I suggested earlier, it is quite doubtful that the evolutionary technique would ever be useful. If the mechanism of consciousness is a redundant system, an evolutionary technique that relies on computers must find suitable functions by trying out infinite possibilities. The author believes that the mechanism of consciousness is a kind of redundant system. I know that the evolutionary technique is excellent as already described, but it is impossible to find the mechanism of consciousness using this technique. The evolutionary technique may be effectively used once the mechanism of consciousness is found.

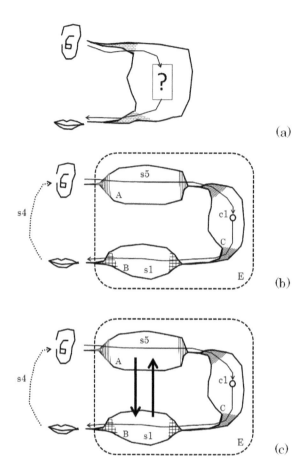

Figure 9.5. Where does consciousness exist?

Roboticists have been using a behavioral model simulating the human brain (Fig. 9.5b). This is generally called a mechatronics model.

In Fig. 9.5b, area A is the recognition system to listen to and understand the content of voice input into the ears. Area C is the decision system to select an action for the language recognized. The ears listen to the word *apple*, for example, and the information reaches c1 in area C via a path s5, where c1 is the language label for *apple*. The language label, also called a representation, is a name given to a certain concept. Area C decides a certain behavior based on the information of language label c1. Signals of the decision run through path s1 in area B, and speech (the corresponding behavior) is implemented by the mouth. Area B is the behavior system and area E the memory area. It is known that s5 and s1 are capable of self-organized (Kohonen, 1995) Note that Fig. 9.5b is insufficient to describe the various known functions of consciousness.

The recognition and behavior systems are directly related to each other as the discovery of mirror neurons shows.

When s5 learns, the neural networks in area B are affected, and when s1 behaves, the neural networks in area A are also affected. This can be explained as shown in Fig. 9.5c. Specifically, when area A learns the function to recognize the word *apple*, area B learns to speak the word *apple* at the same time.

Mirror neurons are neural networks related to both the recognition and behavior systems simultaneously. Accordingly, the recognition and behavior systems may well be defined as shown in Fig. 9.6, which is more direct. Neural networks in area D are related to both area A and B. We cannot immediately determine that area D is the home of the mirror neurons, but undoubtedly this is the area where the functions of mirror neurons are active. The role of area D is for area A (the recognition area) to directly activate area B when area A is activated by learning or recognition, or without using other alternative paths.

Area D is also a shortcut from area B (the behavior system) to area A (the recognition system) by which learning or behavior in area B directly activates area A.

The author defines the source of consciousness to be the coincidence or consistency of cognition and behavior. "Being conscious" includes the state of subconscious according to the author. "Becoming conscious" refers to a state in which any one of multiple subconscious events is lifted to the level of consciousness (i.e., made explicit) as a result of certain conditions being satisfied.

As a human grows and develops, all of his or her functions are learned, and all subconscious events are active, during the state of consistency of cognition and behavior.

For example, the circulation of information through area D and c1 is the state of consistency of cognition and behavior, and the language label c1 is "being conscious."

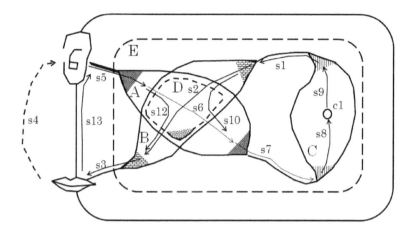

Figure 9.6. Consciousness system proposed by the author.

In Fig. 9.6, c1 is depicted as a single neuron. This is a symbolic expression, and c1 is actually a result of coding. There should be no problem if the expression is spread into multiple neurons. Normally, the binding problem arises in which the representation is expressed by the firing of multiple neurons. In the author's model, the source of consciousness is defined as circulating information, so that it is possible to interpret that signal s8 fires multiple neurons in area C. According to certain conditions, a single subconscious event is selected and surfaces to consciousness, i.e., becomes conscious.

When the consistency of cognition and behavior is lost, the current behavior is immediately stopped and another behavior (relearning) is consciously performed to restore the consistency of cognition and behavior.

With the generation of consciousness defined as above, human mental activities can be explained relatively clearly.

Circulation p1 (s7, s8, s9, s1, s10) does not include path s2. This means that speech is not performed when using this path. This state is interpreted as "imagining" the speech of the word *apple* after recognizing it by hearing. Circulation p1 therefore explains the function of human imagination.

The function of imagination can describe the existence of qualia because the recognition and behavior systems are intertwined in area D and both recognition and behavior information of c1 can fire simultaneously.

On imagining the word *apple*, the system automatically remembers the behavior of speech of *apple*, cognition that the system has heard the word *apple*, and co-occurring memories of *apple*. Co-occurring memory here is a sort of association memory. Co-occurrence means that information circulates through a closed loop with all the neurons in the loop firing simultaneously.

Path s12 in Fig. 9.6 is used to directly transmit the result of the recognition system to the behavior system, bypassing the consciousness proper. This path explains conditional reflexes as well as behaviors directly triggered by a behavior pattern obtained by vision.

An example of a visual-related pattern is a number of small children running in the same direction. When a child totters and falls forward, the children behind follow suit and fall nearly at the same time.

When one is watching a movie and an actor suddenly kicks something with his feet, one suddenly does the same action. This phenomenon is usually suppressed strongly, and seldom occurs in the everyday world. According to psychological knowledge, this phenomenon is known in the medical field and is called imitation behavior (Fig. 9.3).

Consider the case of the patient with partial damage to the frontal lobe, mentioned in Section 9.1.5. When the doctor points at something with a finger, the patient imitates the doctor's gesture unknowingly (Fig. 9.3, top). A pair of glasses is placed in front of both the doctor and the patient. The doctor reaches out his hand to pick up the glasses. The patient, seeing the doctor's movement, immediately performs the same act. The patient is already wearing glasses; yet he was going to pick up and put on another pair of glasses. The patient was instructed by the doctor beforehand not to do anything and the patient was conscious that he should do nothing. Nevertheless, the patient could not stop imitating the doctor (Fig. 9.3, bottom). This case illustrates the existence of information path s12 in Fig. 9.3, through which visual images are directly connected to behavior.

Path s13 transmits the result of the behavior system directly to the recognition system. This explains the phenomenon of self-awareness, i.e., knowing the behavior of the self.

9.2.5 *Consciousness Is Generated by Consistency of Cognition and Behavior*

The author regards imitation as one of the important functions of human consciousness. Imitation is the capability to behave in the same way as one sees by just looking. The case of an infant sticking out the tongue has been already discussed. There are many other examples, including the case where a mother talks to her child, "Tasty, tasty" while feeding the child, and then the child is soon able to speak, "Tasty, tasty," by repeating the mother's words. The child's words may not be very refined at first, but there is no problem because improvements will soon be seen some time after. Imitation is not unique to humans but is found in relatively higher-level animals such as chimpanzees and gorillas. It is marvelous that small children acquire the ability to imitate in a short time. Mitsuo Kawato at the Advanced Telecommunications Research Institute attempted to achieve this ability in a machine (Kawato, 1996).

The author's computational explanation of imitation resembles that of Kawato. The difference lies in that the author describes imitation comprehensively, focusing on the function of consciousness he proposes. An example of the auditory sense, which is different from imitation, is used in the author's description of imitation, but the generality will not be lost even if the auditory sense is used to explain behavioral imitation.

The author describes the function of consciousness referring to Fig. 9.6. The information on the words spoken by the mother flows through paths s5, s6, and s7, thereby activating the neurons of the recognition system A, including the common area D. The information further runs through path s8 and reaches and fires language label c1, which is thereby linked to the activation information of the common area D. Note that the activation information of the common area D is also part of the behavior system B, which is the speech system in the present example. When the consciousness system decides to execute the contents of language label c1, the relevant information c2 flows through s9, s1, s2, and s3, and the contents of language label c1 are spoken. The spoken language label c1 may or may not agree with the mother's words, but when they do, learning is successful.

When the words disagree, the child (i.e., the consciousness system) refers to the speech (the output) signal s4, which the child hears with its ears. The information of s4 goes to s5 and s6, where the value of the common area D is rewritten to correct the difference. Rewriting is repeated until the information circulation s9, s1, s2, s3, s4, s5, s6, s7, s8, and c1 no longer needs rewriting of the common area D.

The final state achieved is the consistency of cognition and behavior.

When this state is achieved, a three-layer structure is established: the external environment, the areas responsible for accepting stimuli and emitting response to external world (area A and B including area D), and the language labeling area C. The author calls this three-layer structure the Consciousness Module. Area C is called the Representation Area. The author believes that the flow of information circulating through external environment, area D and area C is the source of consciousness.

When speech is spoken consciously, the relevant information circulates through c1, s9, s1, s2, s3, s4, s5, s6, s7, s8, and c1. This route indicates that "the self is conscious of speaking the words represented in c1."

The problem is how to create common area D.

Area D must encompass both area A, the recognition system, and area B, the behavior system. Area D must function with regard to both recognition and behavior, and the information recognized must be directly used to perform a behavior. By stating this I must be ready to be impeached by the reader for reason of committing a category mistake. I admit that this is a sort of category error. However, a category is originally a philosophical idea devised by humans. Look at the brain in all of its details, and we just find neurons. There is no area that befits the term category. What exists in area D is artificial neural networks, and it is the human that categorizes it.

The author further believes that common area D provides a solution to the symbol grounding problem because the common area D is located between the external environment and language label, and interrelates them.

An example of the sense of hearing has been already described in this chapter. Let us consider another example to describe the author's idea.

Assume that the mother says "good morning" to the family in a room where bright morning sunshine is shining in through the window, and the smell of soup and bread hangs in the air. If this situation is known to the consciousness system by learning in the past, the consciousness system for visual information and smell information, respectively, will function, and at the same time, their respective common area D will be connected at the representation area C as co-occurring memory. As a result, when "good morning" is given to the perception system, the rest — bright morning sunshine and the smell of soup — is automatically called up. This co-occurring memory also fires the common area D automatically. The author believes that this is a series of so-called association memories. The smell of soup is memorized in the common area D as a result of actually recognizing it in the past with the somatic sensor. Accordingly, the actual smell of soup is reproduced (becomes conscious) as an information when the smell of soup is associated.

I believe this is qualia.

The sense one receives from qualia is different from, at least, the actual sense. This is because qualia are not a consciousness arising from the solid consistency of cognition and behavior closely related to external circumstances and embodiment but are simply produced from circulation of information called imagination (called circulation of imagination). A typical information circulation route is c_1, s_9, s_1, s_{10}, s_7, s_8, and c_1. Certainly, when the smell of soup is sensed from the external environment, the same information circulation occurs, and the same qualia are generated. Information circulation in this case is s_5, s_6, s_7, s_8, c_1, s_9, s_1, s_{10}, s_7, s_8, c_1, and s_9 in Fig. 9.6 by replacing the sense of hearing with the sense of smell. When the smell of soup is quite different from the language label c_1, the system learns to construct a new language label c_1. If the difference is small, the current language label c_1 remains unchanged.

The author believes that the most important function of human consciousness is distinction between the self and others. This is because the distinction of the self from others is the starting point of implementing imitation described in the preceding chapter. Here is a question: How can I know that the hand I am now looking at is my hand?

Figure 9.7.   The self's hand H and the other's hand H1.

This is the theme of this chapter. The principle of the consciousness system proposed by the author is used to answer this question.

Historically, Descartes was the first thinker who clarified
the existence of the self, as already described. Believers in
phenomenology have studied the theme of the self and others
from philosophical points of view. Merleau-Ponty, among others,
proposed the idea of embodiment to describe the spiritual activities
of humans. According to him, the self cognizes the existence of
others in the external environment by using perception obtained via
embodiment.

How does the author differentiate the self and others in the
consciousness system he proposes?

In Fig. 9.7, we read vision for hearing and hand for mouth. Vision
is signified by N and hands by H and H1.

The self's hand is denoted by H and the other's hand by H1
(Fig. 9.7). Self and oneself essentially have different concepts, but in
our description here, we use self as a generic term. Oneself seems to
refer to the whole body, including the flesh. Let us assume that vision
N looks at the other's hand H1 and its movement. The information
of H1 captured by vision runs through s5 and s6 and reaches the
recognition system A, where the hand is recognized.

This information affects the behavior system B via the common
area D, and the behavior system B immediately tries to imitate the
movement of H1 by interpreting it to be the hand of the other.
Actually, however, the self's hand H does not move because the
action of the behavior system B is suppressed. This suppression is
intrinsic to the consciousness system. The flow of information at this
time is s5, s12, and s3. The same information also flows through s5,

s6, s7, and s8 to fire the language label c1. The firing of c1 is linked to the firing information of the common area D. By linking we mean that the firing of the common area D causes the firing of the language label c1 by co-occurrence.

When experience has been sufficiently learned in the past using the route of circulation of imagination (s9, s1, s10, s7, s8, and c1) already described, the firing of the language label c1 ignites the relevant part of the common area D where the path intersects it.

The information and its circulation strengthen the link between the language label c1 and common area D.

To sum up, the information of the other's hand H1 flows through s5, s6, s7, s8, c1, s9, s1, s10, s7, s8, and c1 and creates a closed loop of information circulation consisting of c1, s9, s1, s10, s7, s8, and c1. We interpret this to mean that the information circulation makes us imagine that the hand H1 is the hand of the self by the existence of the common area D. To imagine means that this consciousness system thinks that H1 is its own hand. This is because the common area D is part of the behavior system B. What happens with the self's hand H? It is driven by the behavior system B. The information flows through the route of c1, s9, s1, s2, and s3.

Information on the movement of the hand is returned to the consciousness system via s13, a somatosensory nerve path, and also via s4, or through the external environment. Information s13 provides a means for the recognition system A to know that the behavior system B has issued behavior information. It is also possible that the other consciousness system detects the change of sensors on the skin following the movement of the hand and inputs the relevant data into the recognition system.

Path s4 means that the movement of the hand is detected by vision. It is discussed as follows. The same discussion applies also to s13.

Information s4 enters the vision and becomes information s5, which reaches the recognition system A. This information passes through recognition system A and becomes s6, which flows through s7 and s8, eventually reaching the language label c1 to ignite it. The common area D and the language label c1 are linked to each other, so that the language label c1 is symbol-grounded via common area D. In summary, when the consciousness system views the self's hand

H, the information circulates through c1, s9, s1, s2, s3, s4, s5, s6, s7, s8, and c1. The same information flows through c1, s9, s1, s10, s7, s8, and c1. The same information also runs simultaneously through c1, s9, s1, s2, s13, s6, s7, s8, and c1 via the somatosensory nerve path.

This is the state of the consciousness system in which the language label c1 and the common area D are firing at the same time, or the recognition and behavior systems are said to be consistent with each other. These three information loops are the root source for the consciousness system to become conscious of the self's hand H. The difference of the self's hand H and the other's hand H1 in the consciousness system is that the consciousness system uses the information routes of s2, s3, and s4 as well as s2 and s13 for the self's hand H but does not use them for the other's hand H1.

I will now explain that it is possible to cognize the image of the self in a mirror (i.e., self-consciousness) as the image of the self by defining that consciousness is generated from the consistency of the recognition and behavior systems.

Humans brought self-consciousness into the consciousness system (the self) as a model of being-in-the-world, and by using self-consciousness as a clue, they then succeeded in capturing the model of the world into their consciousness system. This description suggests the existence of a scheme to understand others by replacing the self with the other. A human can cognize his or her image reflected in a mirror to be the image of the self, but it is difficult to explain why.

When did consciousness emerge in the history of human evolution?

It should be natural to think that the self and others were first differentiated from each other due to the necessity of sharing work while living as a group, and then humans became conscious of the self among the others. If it is possible to become conscious of the self in the midst of others, it should also be possible to become conscious of others in relation to the self. To become conscious of the self in the midst of others, a model of consciousness of both the self and others must be created in the consciousness of the self.

Even non-Cartesians may not be able to deny that the function to become conscious of the self and others exists in the consciousness of the self. This is true even if this function consists eventually of reaction formation. I, too, do not deny the structure of reaction formation. I support the idea that any mechanism present in the brain depends on material and information reaction. Of course, the reaction system must be able to describe the mechanism of human consciousness.

In Fig. 9.8, when the representation area C instructs the execution of a hand movement, e.g., swing the hand vertically (h), it transmits the relevant instructions to the behavior system B of the consciousness module G via s1. The behavior system B moves the system H of the hand via s3 to execute the instruction. At this time, the behavior system B of the consciousness module G activates the recognition system A of the consciousness module G via s13, and the action h of the hand is reported back to part of the representation area C via s7. The circulation route of s1, s13, s7, and C is activated repeatedly to transmit information on the moving hand to part of the representation area C. This repetitively circulating information may be said to be the source of being conscious (L) of the fact that the system is moving a hand. The consciousness that a hand is moving is further strengthened by the information circulating through s1, s3, s4, s5, s7, and C. This explanation also applies to the consciousness module R for moving the feet.

This consciousness model can recognize that the image reflected in a mirror is the image of the self. The image in the mirror knows that the vision system N of the self recognizes the features of the body of the self, and the behavior of waving the hand h′ and moving the foot f′ also appears on the image in the mirror simultaneously. After integrating all of these informational data, it is still impossible to tell whether a person resembling the self is doing the same action, the other person is reflected in the mirror, or the image is that of the self.

In the consciousness model proposed by the author, taking the case of the conscious movement of the hand H as an example, all of the circulations p1 (s1, s10, s7, and C), p2 (s1, s12, s7, and C), and p3 (s1, s3, s4, s5, s7, and C) and the recognition route for the image of the self reflected in the mirror q1 (s1, s3, s4′, s5, s7, C, s1, s10, s7, and C) contribute to fortifying the source of consciousness. The consciousness generated by circulations p1, p2, and p3 agrees

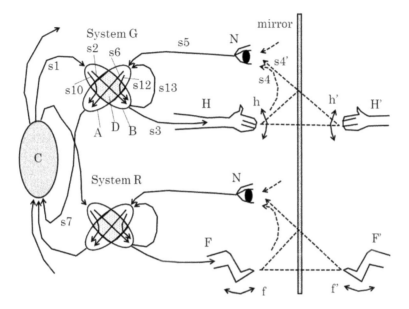

Figure 9.8. Scheme of recognition of self reflected in a mirror.

with the condition of the consistency of behavior and cognition, or that none other than the self is established. A conclusion is also reached that s4 and s4′ are indistinguishable existences since the route q1 cannot be differentiated from circulation p3 (Fig. 9.8) and no reasonable grounds exist to assume that s4 and s4′ are different.

A further investigation is required to prove this indistinguishable existence and to clarify the mathematical relationship between the image of the self and the self. This is because, for example, a mirror does not necessarily reflect an object perfectly. This will be discussed in detail later.

Here, I have introduced the basic procedure for judging that the image of the self reflected in a mirror is none other than the self.

As the functions of higher-level cognition develop, humans will reach the recognition that the image of the self in a mirror is a copy of the self since it is a mathematical existence indistinguishable from the self, and because of the absence of any physical bonding with the body of self.

One will cognize that the reaction of the routes p1, p2, and p3 is on this side of the mirror. This is one way of defining self-consciousness, i.e., becoming conscious of the self.

The circulations p1, p2, and p3 and the route q1 are a tool to interpret one's own body and external environment simultaneously, and also a powerful tool to take in the external environment by learning.

According to the consciousness system developed by the author, p3 and q1 are two different routes that show the same reaction with regard to consciousness. Using this feature, route q1 can be taken into the representation area. This means that multiple models of the self can be taken into the representation area one after another. Using this scheme, the representation area builds models of the self and creates self-consciousness in the consciousness system. Consciousness of others is taken into the representation area with a similar procedure by comparing it with the consciousness of the self. Once consciousness of the self and others has been taken into the representation area, it is possible to introspect one's own feelings and become thoughtful of the feelings of others. Mutual internal and external communications are also possible once consciousness of the self and others is taken into the representation area. This process provides a means for capturing being-in-the-world into the representation area.

9.3 My Standpoint

9.3.1 New Materialism

The author is basically a materialist and, as such, believes that the human brain can be analyzed as a substance. It is necessary, however, to include the concept of information in the category of substance. Conventionally, information, together with energy, is categorized as different from substance. Energy and substance can be expressed mathematically by expressions (e.g., $E = MC^2$). The relationship between information and substance may also be expressed by a numerical expression in the future.

Information is accumulated in a substance, and only then can information affect the world environment using the substance. It may sound extreme, but when two different pieces of information, a and b, are present in the same substance N, can't we say there are two different substances Na and Nb because they can have different functions to interact with the external world? By thinking in this way, one may consider the flow of information through substances to be a concept of a new materialism. Computers have traditionally been treated materialistically as an information processor, and therefore the above description by the author may not necessarily be unnatural. People often say, "Consciousness and the mind do not exist. They are but an illusion or a reverie," when discussing whether consciousness and the mind exist. This is a confusion arising from the old materialism fettered by the doctrine that consciousness and the mind do not exist as substances.

9.3.2 *Connectionism*

The author's standpoint is connectionism. Connectionists believe that the human brain is made up of networks of neurons and their functions create the spirituality of a person. They further believe that artificial consciousness can be generated by connecting artificial neural networks.

9.3.3 *Cognitism and Representation*

The author sides with cognitism and acknowledges the existence of representation. Representation means images of external and internal objects occurring in the consciousness based on perception, i.e., subjective concepts in the brain corresponding to objective things existent outside the body or in the internal world. This idea was inaugurated by German philosopher Immanuel Kant (1724–1804) (Kant, *Critique of Pure Reason*). Simply put, when you see an apple, a recognition that it is an apple is formed in your brain. This means that a subjective image of an apple is created with respect to the objective thing of an apple present in the external world. The subjective image in the brain is called the representation of an apple. We acknowledge the existence of representation because it would be difficult to describe the commonest phenomenon of human speech if representation were denied.

9.3.4 *Not Emergent*

The author does not agree with the idea of explaining all about consciousness by emergence. Emergentism is derived from behaviorism. Robots developed by behaviorists are considered to have a set of simple behavior protocols that are selectively implemented in response to changes in the environment. This idea is understandable, but how is the high-level cognition function unique to humans created in their robots? How is human thinking created in them?

Behaviorists use the word "emergent" when such questions are asked. Emergent means that during the process of evolution of high-level cognition function of humans, certain abilities or features are obtained by chance. The author has no objection to the idea that the function of cognition was acquired through evolution. However, even though the mechanism is unknown, no one should negate scientists' efforts to estimate and define the principle of the mechanism. Emergentists' research technique is to leave a robot in a difficult situation and expect it to solve the problem and evolve by itself (evolutional robotics). This may be a feasible research technique, but it would be difficult for the robot to discover the function of consciousness through such a process. This evolutionary technique is not effective if the function of consciousness is a redundancy system.

9.3.5 *Avoiding Infinite Retreat*

The third property of consciousness by Husserl reads that "the self becomes aware of the self." The word "self" is used twice in this statement. This is the same as saying that there exists a self who becomes aware of the self. This further leads to "there exists a self who becomes aware that 'there exists a self who becomes aware of the self,'" and the self exists eternally while retreating (Fig. 9.9). If this statement is accepted, the function of consciousness cannot be described by computers because computers cannot solve a problem including a recursive retreat and therefore cannot end the computation process.

Using this fact, a researcher has declared that the human brain includes non-computational elements (Penrose, 1994).

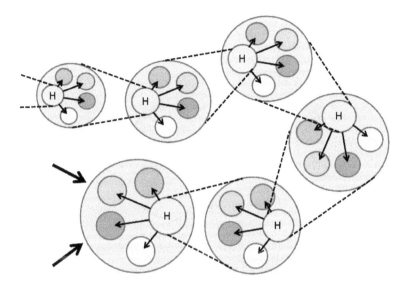

Figure 9.9. Recursive retreat.

The author believes that the recursive retreat problem can be somehow avoided by using infinite repetition (Fig. 9.10). The state of becoming aware of the self is understandable, but it is impossible to actually feel the self who retreats multiple times in the realm of phenomenology. I believe that the self who retreats multiple times is simply an imaginary object of symbolic logic created in the brain because we are already handling the concept of infinity as a symbol. The author, therefore, believes that it suffices if we can create a self who becomes conscious of the self on a computer, and the rest is considered to be the result of symbolic logic within the brain.

9.3.6 *Quantum Consciousness*

Quantum theory would provide an effective means of research if we were to analyze the brain at the atomic level. There are many things we can do, however, before we resort to quantum theory. I believe that quantum theory is not necessary if one is interested in elucidating the mechanism of consciousness.

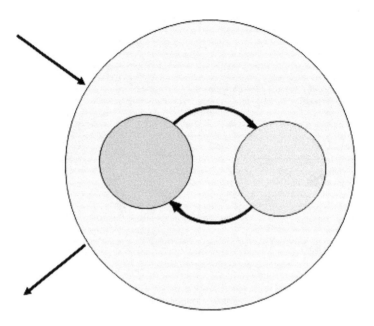

Figure 9.10. Infinite repetition.

9.4 Computational Model of the Consciousness Module

This chapter describes the MoNAD, the consciousness module that is the basis of the consciousness system.

9.4.1 *Overview of Computation Model*

A MoNAD is an artificial neural network with a slightly complex structure that is built upon the concept model of consciousness described in the previous chapter (Fig. 9.11). The first feature of the MoNAD is that at least two closed loops (c1 and c2) exist and both are connected to a common neuron group K called primary representation. The second feature is the somatosensory nerves Y' extending in both common neuron group K and the closed loop c2. The MoNAD has at least one input terminal Input L. Closed loop c1 has two symbolic representations Z and Z'. Z is called cognitive representation and Z' behavioral representation. A MoNAD learns the neural networks so that the values of Z and Z' are equivalent.

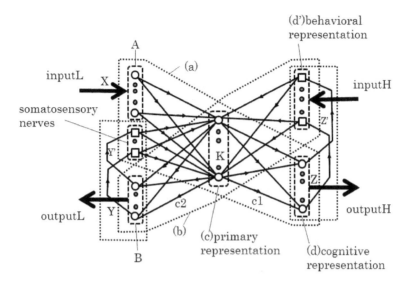

Figure 9.11. Computational model of consciousness module.

When a behavioral representation Z′ is decided, the MoNAD executes behavior Y from output L, and the information X obtained from input L is changed to make the value of Z coincide with that of Z′. Z and Z′ also serve as I/O terminals (output H and input H, respectively) for exchanging information with higher-level MoNADs. Information on the current behavior is input again into K via somatosensory nerve Y′. The MoNAD, thus, cognizes not only the external environment but also the behavior of the self. The result of this cognition is reflected on the information of Z, and this information represents both the external environment and the behavior of the self.

9.4.2 *MoNAD Functions*

The MoNAD is designed to satisfy the consistency of cognition and behavior, which is the author's definition of consciousness. This means that in the MoNAD information runs consistently through the neural networks. In this model, the information flows through Z′ of the closed loop c1 and common neuron group K, and then reaches Z. The value of Z is copied to Z′ and is also input into the input L of a higher-level MoNAD. Information from Y runs through

Y′ of the closed loop c2 and K and then returns to Y. Information basically circulates in a single MoNAD through K, Z, Z′ and returns to K, from which the information further flows through Y and Y′ and then returns to K. Information X from input L is added to this circulating information all the time. If the internal circulation of information is maintained normally despite the input of incessantly changing external information X, the MoNAD is said to be keeping the "consistency of cognition and behavior." This is because the information flow of (Z′, X, Y′), K, Y decides the next behavior to be taken and the information flow of (Z′, X, Y′), K, Z decides cognition. Information from Y output from the terminal output L is also a new order for the lower-level MoNADs to engage in a behavior because Y changes the behavior of the lower-level MoNADs. When the value of Z′ is changed by a higher-level module, the MoNAD executes an action from output L that will eventually make the value of Z coincide with the value of Z′. The author calls this state, in which the value of Z is equal to that of Z′, a state in which cognition and behavior are consistent.

9.4.3 *Investigating Husserl's 10 Properties*

Husserl enumerated 10 properties of consciousness. Does the MoNAD proposed by the author satisfy them?

(1) The first-person property is the assertion of the existence of the self based on mind–body monism. The representation area of the MoNAD distinguishes between the self and others and represents the behavior of both. This means that the MoNAD recognizes the behavior of others based on the behavior of the self. This is a sort of mind–body monism.

(2) Orientation means that consciousness is always directed toward something. The MoNAD self-learns to behave, by output L, to make cognitive representation Z coincide with behavioral representation Z′. As such, the MoNAD satisfies the second requirement of Husserl.

(3) This is the same as point 2. In addition, the duality of consciousness is that the self becomes aware of the self. The MoNAD is capable of cognizing the behavior of the self using

its somatic sensation Y′. The MoNAD is therefore able to become aware of the behavior of the self by itself.

(4) Expectation: The MoNAD performs a behavior (output L) according to point 2 such that the cognitive representation Z agrees with the behavioral representation Z′. This means that the behavioral representation Z′ expects the forthcoming cognitive and behavioral state of the MoNAD.

(5) Conviction of the existence of things: The MoNAD always calculates the cognitive representation Z using external information (input L) and internal somatosensory information Y′, .i.e., Z results from the cognition of the external and internal environments. In the area of the MoNAD where external information is directly processed, information X about the external environment inevitably includes uncertainty but Z, as a representation, always converges on a certain definite value. The MoNAD can, thus, discern a positive existence in the midst of uncertain information.

(6) The body is part of the self: The MoNAD cognizes the behavior of the self using its somatic sensations. This means that the MoNAD cognizes and is convinced that the present behavior is being done by the self, i.e., the MoNAD is confident that the behavior of the self is being done by the body of the self.

(7) Distinction between the self and others: As stated in point 6, the MoNAD continually cognizes things by obtaining somatosensory information and information about the external environment via input L. This means that the MoNAD always cognizes information about the behavior of the self and the external environment. As such, I conclude that the MoNAD distinguishes between the self and others.

(8) Property in which reason is related to emotion: With the properties 5 through 7, the MoNAD has functions to represent the internal and external environments. If this is true, the MoNAD can represent information about emotions arising in the body. Representation is shown on a Pleasant or an Unpleasant MoNAD. This means that reason and emotion can be achieved as a representation function using MoNADs. Details will be described later. As Antonio Damasio,

Professor at the University of Southern California, says in his book on Baruch De Spinoza (1632–1677), humans seem to have developed in such a way that reason and emotion function together (Damasio, 2003). However, his book does not discuss the feelings of the body arising from emotions.

(9) Randomness: The MoNAD is made of artificial neural networks. For this reason, networks are always swaying or in a state of mathematical chaos. A consciousness system structured using layered MoNADs can define the sway of consciousness, but we will not go any further into this topic right now.

(10) The qualia problem: The MoNAD does not define its relationship with qualia in detail. This will be discussed later.

9.5 Discriminating the Self from Others Using the Function of Visual Imitation

The MoNAD is capable of learning the function of imitation. Imitation means that the value Y of output L follows the value X of input L in the MoNAD. In principle, imitation means that the input value X of the external environment is output as it is to Y (and simultaneously to Y′). The MoNAD cognizes the external and internal environments and indicates the result on the cognitive representation Z. This means that the MoNAD distinguishes between the self and others. When the external and internal states are in agreement with each other, imitation is successful.

If the cognitive representation Z and behavioral representation Z′ are in agreement, cognition and behavior are consistent in the MoNAD. Irrespective of whether they are in agreement or not, the value Z is copied to Z′. This means that the MoNAD basically accepts the condition of the real world.

9.6 Solution of the Symbol Grounding Problem

A computer is basically a high-speed symbol processor. The problem is that the symbols processed inside are semantically disconnected from the external world. As I mentioned before, this problem is the major cause of the stalemate that conventional artificial intelligence (AI) is experiencing. The MoNAD possesses the function to represent the external and internal environments simultaneously. Since this representation is a symbol in a sense, the MoNAD continues to interpret the real external and internal worlds as symbols. Because of this property, the MoNAD may be considered to have the function to solve the symbol grounding problem.

9.7 Consciousness System

The consciousness system shown in Fig. 9.12 consists of layered MoNADs. Each MoNAD is driven by the top-down and bottom-up system to continually exchange information between the lower- and higher-level MoNADs. As the basic flow, each MoNAD executes a behavior (output L) based on information from the lower- and higher-level MoNADs (input L and input H, respectively) and the somatosensory information Y′. This information is constantly

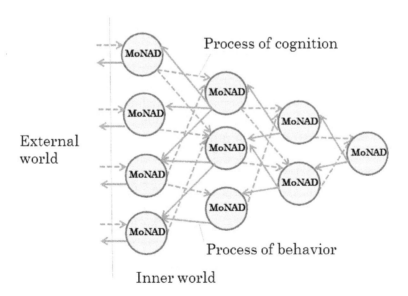

Figure 9.12. Consciousness system structured using layered MoNADs.

transmitted to the higher-level MoNADs via output H. Information Z is normally copied as is to Z′ except that when new information is given to Z′ from a higher-level MoNAD via input H, the receiving MoNAD works to create Z′.

The lowest-level MoNAD is responsible for communicating information with the external world. Behaviors or transactions with the external world are implemented via output L, and the external information enters the MoNAD via input L. Excepting the transfer of external information, the MoNADs communicate with one another using their respective representations as information sources. The consciousness system is, thus, a representation-driven mechanism.

The functions of the consciousxness system are described as follows. The consciousness system performs the behavior described below in response to external stimuli while considering its current internal state. What makes this robot different from conventional ones is this consideration of its internal state. The internal state means the whole history of somatosensory information and behavioral representations experienced by the robot up to the present.

Each MoNAD functions independently in the absence of a behavioral representation from a higher-level MoNAD. If a behavioral representation exists, the higher- and lower-level MoNADs function cooperatively. Each MoNAD continually interprets its current status while obeying instructions from higher-level MoNADs. Note that these higher-level MoNADs do not constitute a central control tower known as a Homunculus because the behavior of a higher-level MoNAD is dependent on information from a lower-level MoNAD. As such, it can be said that higher-level MoNADs follow the commands of lower-level MoNADs. Normally, representations of the higher-level MoNADs are concepts of a higher level compared with those of the lower-level MoNADs.

The current system, including its constituent MoNADs, uses a human-supervised backpropagation learning method. Voluntary development and learning are expected in the future.

9.7.1 *Subconsciousness and Explicit Consciousness*

Humans seem to always have multiple flows of consciousness, most of which are normally not known to them and are thus called the subconscious. You can walk while eating ice cream, for example. It is possible for you to be conscious of eating and not to be conscious of walking. Without being conscious of walking, you can accomplish the other behavior of eating without inviting an accident. Walking is an act of the subconscious, whereas eating is an act of explicit consciousness. The reverse is also feasible. Humans seem to be conscious of just one thing at any given time. What is the mechanism that arouses the subconscious to consciousness? The author assumes that the cognitive and behavioral circulation involving the MoNADs that associate (i.e., connect) the reason and feelings systems (see the next chapter) present in many streams of subconscious are related to explicit consciousness.

This suggests that the actualization of the subconscious is deeply related to emotions. Any subconscious activity producing stronger emotions is considered to be actualized. In the example concerning walking while eating ice cream, eating ice cream arouses stronger emotions than walking, and that is why the consciousness of eating is made explicit. If the balance of the body were to be lost due to stumbling on a stone, the relevant information triggers other emotions and awakens new consciousness. Generally, the subconscious includes emotions excited by the subdued and explicit consciousness. When consciousness is aroused in a human, both the emotion and reason systems of consciousness are working together, i.e., the explicit consciousness refers to a state in which the self is feeling its body emotionally. Assume, for example, that the representation of a feeling MoNAD turns unpleasant in the consciousness model developed by the author. The unpleasant information is transferred to the association MoNAD. The association MoNAD activates the MoNADs of the reason system to engage in a behavior intended to change the unpleasant information to a pleasant one. At this time, the consciousness system is explicit. Explicit consciousness refers to a state in which the representation of the self and the representation of feelings co-occur and react.

9.7.2 *Relationship Between Reason and Feelings*

Humans possess both reason and feelings. No one can deny this fact. Antonio Damasio proposes the somatic marker hypothesis, stating that reason and feelings are inseparable in humans (Damasio, 1999). The author shares the same idea. How are reason and feelings of humans related to each other?

Joseph E. LeDoux, professor of neuroscience and psychology at New York University, has published many achievements in the study of emotion and behaviors (Ledoux, 1996, 2002).

The author proposes a MoNAD model to explain the relationship between reason and feelings. Figure 9.13 illustrates how the most basic MoNAD relates with feelings. Assuming that human feelings are essentially represented by Pleasant and Unpleasant information, the author devised their respective MoNADs D and E (Fig. 9.13). Considering that feelings arise from emotions present in various parts of the body, the author allocated MoNADs B and C to represent the emotion area. MoNAD B, for example, represents the emotion

Reason system

Emotion & feeling system

Figure 9.13. Relationship between reason and feelings.

of "pain" and MoNAD C, "emptiness" due to the loss of values from the vision sensors. The emotions B and C transmit their information to MoNADs D and E, which represent Pleasant and Unpleasant, respectively (bd, be, cd, and ce). The representation of the feelings of Pain and Emptiness is normally Unpleasant, but in this model, information is also transmitted to the MoNAD of Pleasant (bd and cd). Both the reason and feelings systems are constructed using MoNADs. MoNAD F associates (i.e., combines) the information of both the reason and feelings systems. MoNAD F receives information (af, df, and ef) from the lower-level MoNADs (A, D, and E in Fig. 9.13), generates new information (fa, fd, and fe) based on the information received, and transmits it to the lower-level MoNADs (A, D, and E in Fig. 9.13).

This new information means that when the representation af, df, and ef reads the "current cognitive representation is Unpleasant," information fa, fd, and fe is sent to MoNAD A of the reason system urging it to initiate a behavior to change Unpleasant to Pleasant. This means that the information expects that the representation D will be changed to Pleasant.

A system in which the representation of MoNAD D becomes Pleasant if behavior a' is initiated when the external inputs are a, b, and c has already been empirically learned.

In the example shown in Fig. 9.13, the neural circuitry network G judges the coincidence of the behavioral and cognitive representations. Information ad and ae makes the feeling representation Pleasant if the expectation information coincides with the real-world cognitive information, and makes it Unpleasant otherwise. I believe this example is very generality.

The problem here is the development through the learning of the MoNAD that associates (i.e., connects) information. The system developed by the author evolves by repeating empirical learning, and at least in the initial stage of evolution, wiring is expected to have been intrinsically set. Each MoNAD has at least two pairs of bidirectional communication lines to support creation of a new representation by associating simultaneously occurring representations. Since plural representations are associated to a single representation, this model can explain the phenomenon of creativity or the occurrence of a new concept.

Silvano Arieti (born in Italy, 1914–1981), professor of psychiatry at New York Medical School, left many famous research studies on the mechanism of creation in the brain (Arieti, 1976). There is no significant difference between his description and the creativity of the MoNAD developed by the author, but a detailed discussion is omitted here.

My discussion of reason and feelings continues. Changes in the behavioral representation MoNADs D and E are transmitted to the lower-level MoNADs B and C via db, dc, eb, and ec to change the behavioral representations of these MoNADs. This indicates that the emotional MoNADs B and C are activated when the system cognizes the representations Pleasant and Unpleasant. The emotional MoNADs B and C, respectively, create new representations by adding information from by external stimuli (b, c), and then send those representations to the higher-level MoNADs D and E to change their representations. This means that the system operates in a way similar to a human's changing to a state of Pleasant or Unpleasant just by thinking of Pleasant or Unpleasant.

The consciousness system with emotions shown in Fig. 9.13 consists of three essential MoNAD systems: reason, feelings, and association. The reason system executes behaviors in the external environment and at the same time interprets and represents the external environment. The feelings system represents the information of the emotion system activated by external stimuli and also represents the feelings aroused by the information represented in the reason system. The association system transmits the information of the feelings system to the reason system and also sends the information of the reason system to the feelings system. Generally, there is the major circulation of information through the feelings, association, reason systems, and then back to the feelings system. There are, theoretically, direct inputs from feelings to the MoNADs of the reason system (da, ea), but a description of these inputs is omitted here. Information from feelings always passes through the association system. The reason system executes behaviors to interact with and change the external world. As a result of these behaviors, the external information is changed and the internal representation modified. Importantly, in the consciousness system incorporating the MoNADs proposed by the author, the internal environment is also variable with the changing condition of the somatosensory neural networks. In the author's system, the next behavior is decided not only by depending on the condition of the external environment but also by considering the condition of the internal environment, including feelings. The reason system that interprets the external environment is capable of interpreting it emotionally to deliver information to the feelings system. This means that the system has a means to evaluate the feelings of others existing in the external world.

The operation of the emotion system is described as follows.

The internal and external information is required to change the representations of the feelings MoNADs. Internal information includes reflex information from the association system and information derived from the representation of the reason system. The section using the neural network G is currently mounted on the robot. The section using the representation of the reason system itself is a task for the future. The model shown in Fig. 9.13 describes the relationship between the MoNAD and the feelings system. It is clear that the consciousness system incorporating these MoNADs is capable of expressing Pleasant and Unpleasant feelings. A system in which more than one MoNAD exists in the reason system is yet to be studied.

9.8 Where Do Qualia Come From?

The reader may claim that the MoNADs described here lack the function to explain qualia. That's right. When the author earlier discussed the relationship between qualia and the somatosensory system, he said that the functions of consciousness and qualia are different problems as far as the MoNADs are concerned. Qualia have been considered the hardest of all problems in the function of consciousness (http://consc.net/papers/facing.html). The author assumes that qualia are a function below the function of consciousness from the fact that "to become conscious of feeling something" is easily understood, whereas "to feel to become conscious of something" is not. I agree that the qualia problem is an interesting theme for scientists, but if qualia are not directly related to the function of consciousness, it would be better for now to stay away from the qualia problem. If qualia are a subjective yet physical phenomenon, someone will solve this problem in the future, and at such a time, the function of consciousness and qualia might be combined. The author's function of consciousness is designed to allow the incorporation of a qualia function later. In brief, the function of consciousness can cognize the qualia present at a given time consistently because the author defines the function of consciousness to be consistent cognition of the behaviors of the self and others.

The author believes that the mystery of the occurrence of qualia is related to the transformation of information into substances and vice versa. At the end of each nerve, information is generated from substances, and in the brain, substances are produced from information. Transformation occurs cyclically in both the body and the brain using bidirectional information transmission routes via the MoNADs. This is the idea that any event being experienced in the body is virtually experienced in the brain at the same time. This is synonymous to the old idea of a bodily model of the self or a homunculus in the brain. Anyway, the details are yet to be clarified.

I would like to introduce my view of qualia.

The function of consciousness has been substantiated scientifically by the author on the basis of computation theory, but qualia still remain a subjective problem. Qualia are a sense of the self and cannot be understood by the sense of others. For example, it is impossible to prove if my pain and my wife's pain are of the same sense. This question is asked by Thomas Nagel in his essay, "What Is It Like to Be a Bat?" (http://members.aol.com/NeoNoetics/Nagel_Bat.html)

I think we can believe for now that my pain and my wife's pain are of the same sense. A scientist should not use the word "believe" in this context, but isn't it true that we live by believing? Just because we believe that my wife's pain is the same as my pain, we try hard to remove the pain that brings about Unpleasant feelings from my wife. The quality and severity of the pain, as well as the interpretation of Pleasant and Unpleasant, may vary slightly depending on the people owing to innate propensity and developmental learning, but the essential points should be the same for all people. If not, humans must have lost an effective means of communication and would have become extinct long ago in the process of evolution. While acknowledging the existence of small differences, we should judge that there is no particularly significant difference.

Consider the exquisite music composed by Beethoven and Mozart, the graceful Parthenon temple in Greece, and the beautiful balance of the golden proportion. All these achievements are appreciated by a multitude of people. From this fact, can't we say that humans have nearly identical qualia? What, then, is the difference between humans and bats? Although both of them are mammals, humans can never feel that they are "the same as bats." The basis of qualia may possibly have been the same for humans and bats at the outset because they were born and have evolved in the same environment of the Earth. We should judge, however, that our qualia are different from those of bats because humans and bats followed different paths of development thereafter.

The author understands that every living body possesses qualia from the time it is born. The author further believes that qualia are a phenomenon captured by the function of consciousness in the process of evolution or through developmental learning. The function of consciousness represents internal phenomena as it represents the phenomena of the external world. One of these internal phenomena is qualia.

The function of consciousness interacts with the external world and ascertains its existence by executing behaviors, or else by means of a scientific means to certify the existence of the external world. The difficulty with the qualia problem is the lack of a scientific means to ascertain the existence of qualia that are a phenomenon of the internal world, and as such, there is a possibility of agnosticism gaining power in the discussion of qualia.

Then, how about humans and machines?

Does a machine have qualia? Scientific arguments will go like this: Machines do not have the function to generate or feel qualia, and therefore machines cannot have qualia. Igor Alexander, a British researcher of machine consciousness, claims that MAGNUS, a conscious system he developed, has qualia (Aleksander, 1996). He asserts that machines have their own qualia. Although his argument has not been readily accepted, it can be one of the explanations deserving further investigation if the qualia problem is left for discussion at the mercy of Kant's agnosticism. Of course, agnosticism is a manifestation of a thought, and as such, there can be another thought to deny it. Discussions in this regard are not over. The author currently holds the same views as Alexander.

9.9 Problem of Free Will

The problem of free will was mentioned in Chapter 4. The topic is discussed in this chapter in detail while referring to the model of the consciousness system.

Research on human free will by Benjamin Libet, professor emeritus at the University of California, San Francisco (UCSF), and other researchers highlighted the theme of free will (Libet, 2004). These research studies doubt the common thinking that humans act based on their own free will. New scientific evidence found by the researchers is that the brain is activated about 0.55 sec before a human arbitrarily decides on and performs a behavior. Humans are unaware of the fact that activities regarding a behavior have already begun in the brain, and they become aware just when the behavior is about to be performed.

When humans think, they become aware of the content of the thought in some cases. For this reason, there are two types of cerebral activities by humans: those noticed and those not noticed. Cerebral activities occurring before one becomes aware may be an integrated process of what has been learned empirically all of one's life since one was born and the functions that one had already possessed when one was born. The latter refers to the innate functions acquired in the evolution process of life since life appeared on the Earth. Because these functions are built into and belong to individuals, the cerebral activities are performed with self-responsibility, irrespective of whether one is aware of them or not. This is the basic idea underlying the thought of free will.

Free will means not to receive instructions from others and that all decisions are made by oneself. On the other hand, when one becomes aware of the behavior selected by the cerebral activities, one is given the opportunity to scrutinize the proposed behavior and then perform the behavior. It can be said that one is acting with self-responsibility. We cannot be aware of the scrutiny activity of the brain, but we have the option not to do the behavior once it surfaces to our consciousness. This may be called free will. Wrist-flexing is not a very common social activity; so it may be difficult to use it to discuss the importance of suppressing behavior as it relates to free will. If wrist-flexing were replaced by the gesture of aiming a gun at a person, for example, the role of free will to suppress a behavior would be more keenly appreciated.

In later research, Libet found that an awareness of conscious will occurs about 0.15 to 0.2 sec prior to the physical action, and just at that moment, one can select between stopping the behavior or going ahead with it. Libet formulated a hypothesis that the possibility of this selection is the entity of human will. The brain is always ready to perform behaviors, but consciousness controls the execution of behaviors. In short, humans do have free will and are held responsible for their actions.

The author's model does not specifically use control signals for suppressing the execution of a behavior. However, the author's consciousness system can execute suppressed behaviors without jeopardizing the consistency of cognition and behavior (Suzuki *et al.*, 2005; Takeno *et al.*, 2005) Therefore, it is possible to suppress behaviors using signals to execute or not execute. A suppressed behavior means the act of imagining the behavior. A process seems to exist in the human brain in which the brain imagines a behavior just before the behavior is executed and determines the possibility of releasing the suppression by evaluating the condition of representations firing at that time.

It is possible for the consciousness system proposed by the author to have a similar function to the human mechanism of free will.

The problem here is the relationship between will and responsibility, i.e., who is responsible for the behaviors performed by humans? Generally, if a behavior is performed with awareness, the performer is responsible for the behavior. If, on the other hand, a behavior is performed without awareness or unconsciously, the performer may not be held responsible. And even if the unconscious performer is held responsible, the responsibility seems to be lighter than that assumed for a conscious performer. If a robot has the function of will, can the above concept be applied also to the robot? The problem is, who is responsible for the behavior of a robot? If we make the robot responsible, is there any practical meaning to stating that the robot accepts the blame? If a robot has a self, should it be responsible for its own behavior like a human? The author believes that although robots may have nearly the same functions as humans, the maker of the robot should be basically held responsible for the conduct of the robot, at least for the time being. This would be much like the case of an automobile, in which the owner or driver is held responsible, and various legal regulations should be established to ensure the safe use and operation of robots. Robot owners should also be subject to various legal controls to promote the safe use of robots. We should do our best to ensure the safety of robots so that conscious robots can work in society for the benefit of humans.

9.10 Summary and Observations

This chapter introduced the word's first self-aware robot capable of becoming conscious of itself reflected in a mirror. The scheme of consciousness is made up of recurrent neural networks. The basic idea is as follows: (1) The process for the robot to cognize information coming from sensors and the process to execute behaviors share a common neuron group K, and (2) the common neuron group K is represented by neural circuitry. Another feature of the author's robot is that behavior information is returned to the common neuron group K via somatosensory nerves. The robot thus cognizes information about the external world supplied via the sensors and information on somatic sensation, i.e., both external and internal information, as representations, simultaneously and at all times. This means that the robot distinguishes between the self and others. This scheme of consciousness has been demonstrated to be capable of learning and performing imitation behavior.

Using this scheme of consciousness, the author has constructed emotion, feelings, and association areas in the consciousness system of the robot. The structure, though simple at this stage of development, shows the possibility of the robot's consciousness interacting with emotion and feelings. Subconsciousness and explicit consciousness have been described. Lastly, the sense of a robot, i.e., the problem of qualia, free will, and problems of responsibility related to free will, have also been discussed.

Human consciousness has not yet been completely defined, but the author's scheme satisfies 90% of Husserl's 10 properties of consciousness. As presented in the online version of the Discovery Channel (www.discovery.com), the scheme of consciousness developed by the author is a solution to the problem of mirror image cognition, i.e., becoming aware of the self reflected in a mirror.

Chapter 10

Physical Demonstration of Successful Mirror Image Cognition by a Robot

Humans can be aware of their images reflected in a mirror, but the relevant mechanism — the reason why they can — has never been described physically and mathematically and still remains a mystery. The author has developed a computer program and conducted experiments to achieve awareness in a robot.

Three experiments have been conducted:

(1) A robot imitates the image of itself reflected in a mirror.
(2) A robot imitates another robot of the same type that is made to behave in the same way as the first robot.
(3) Two identical robots with the same functions were placed face to face to imitate each other.

The coincidence of the behavior between the self and the other is calculated in each experiment. The coincidence in Experiment 1 (mirror image test) is always higher than that observed in Experiment 2 (ordered behavior test). In Experiment 2, the other robot that is controlled by the self robot can be regarded as part of the self robot, like a limb in the case of a human. From these experiments, one can say that the image in the mirror is an existence

Creation of a Conscious Robot: Mirror Image Cognition and Self-Awareness
Junichi Takeno
Copyright © 2013 Pan Stanford Publishing Pte. Ltd.
ISBN 978-981-4364-49-2 (Hardcover), 978-981-4364-50-8 (eBook)
www.panstanford.com

that is closer to the self than a part of the self is, i.e., it is none other than the self, and the self is controlling the mirror image. These were the first experiments to physically describe mirror image cognition, and we achieved a 100% rate of success in mirror image cognition with a robot.

This chapter contains detailed logical and physical descriptions of how this result was attained. It also includes observations and prospects derived from the experiments.

10.1 Introduction

Humans can easily identify their images reflected in a mirror despite the fact that since birth no one has ever seen his or her own face directly. This is strange. Humans say, "I know my face by looking at it in a mirror." However, this does not solve the problem. Humans are also said to become aware of their own images reflected in a mirror when they are about two years old (Amsterdam, 1972).

G. Gallup, Jr., proposed a mirror test to evaluate such awareness (Gallup, 1970). So far, chimpanzees, orangutans, dolphins, Indian elephants, and magpies have passed the mirror test.

Jacques Lacan, a French psychoanalyst and philosopher, announced his mirror stage hypothesis, in which he cited the mirror test as an essential milestone in human growth and development (Lacan, 1982).

Becoming aware of one's self image in a mirror is said to suggest the existence of self-consciousness because humans put on makeup and dress themselves in front of a mirror. An explanation of this becoming aware of one's self image in a mirror might possibly lead to a solution of the problem of consciousness.

When we refer to consciousness, we immediately run into a difficult problem. There are groups on two sides: those who do not acknowledge the existence of consciousness and those who do. The former group generally includes scientists on the engineering side, whereas the latter consists of scientists on the physical side. This split is due to a lack of a clear description of the phenomena, i.e., a definition, of consciousness.

The extremists in the former group believe that consciousness is a subjective phenomenon and cannot be described mathematically, and therefore its existence is not acknowledged. Other researchers in the same group also do not acknowledge the existence of consciousness but assume the existence of various functions of recognition, which are integrally combined to form something new, which they call "emergence." For them, consciousness is simply a phantom or an illusion.

The researchers belonging to the latter group acknowledge the existence of consciousness in humans and try to locate it in the structure of the human brain.

Given these circumstances, the effort to discover the scheme of recognizing one's self image in a mirror as a clue to solving the problem of consciousness may well bring on furious arguments between both sides.

The author decided to challenge this task in a way not attempted much by researchers in the past. I decided to define consciousness physically and mathematically by checking research studies on consciousness published in the past. The reason why such an approach had seldom been used in the past is a persistent spiritual resistance to any attempt to describe human consciousness physically and mathematically. This arises from the firm belief that humans are different from machines. Perhaps the remote possibility of actually attaining the aim may be another reason. It is generally believed that a definition of personal consciousness is far from a universal truth.

If the attempt to define consciousness concretely, both physically and mathematically, is postponed at all times, we cannot take even the first step toward our goal of understanding the mechanism of human consciousness. It may be one way of research to expect "emergent" phenomena resembling consciousness by combining various functions of recognition. In the field of brain science, a variety of knowledge may be obtained to prove the repetitive reactions between the brain and the body. Even in these cases, we will still be confronted, eventually, with the difficult problem of defining consciousness. Although we have not yet found a clear-cut and universally acceptable definition of consciousness, we should never give up answering the question, "What is consciousness?" I believe the process I am going to use is an important scientific technique for understanding human consciousness.

Why should we hasten in our efforts to understand human consciousness?

It is naturally great to enjoy the fruits of understanding human consciousness or the mechanism of thought or behavior. I would like to emphasize the following three merits in particular:

1. The contribution to the understanding and discovery of treatment methods for brain diseases, including schizophrenia
2. The development of a means of recovery after the loss of consciousness caused by accidents
3. The development of artificial limbs devised to feel like a person's original limbs for those who have lost limbs in accidents and due to other causes.

I mounted a computer program on an existing small robot. The program was written using architecture developed based on my subjective definition of consciousness. The architecture uses a top-down design based on my original ideas, but the program itself runs using a bottom-up approach. The program not only drives the robot to physically demonstrate mirror image cognition but also describes most of the items related to consciousness that are known today (Takeno *et al.*, 2005).

The author's robot imitates the behavior of another robot placed in front of it according to the built-in program and calculates the rate of coincidence of behaviors between the self and the other. The three important experiments conducted with the robot fully and physically describe the problem of mirror image cognition. This is the meaning of what was described as a 100% successful experiment. The experiment was successfully conducted on September 1, 2004 (Takeno, 2005). The experiment was introduced on the online version of the Discovery Channel 1 TV, the United States, on December 21, 2005 (http://www.rs.cs.meiji.ac.jp/Takeno_Archive/Discovery NewsAwareRobot211205.pdf; http://www.rs.cs.meiji.ac.jp/ Takeno_Archive.html). Now, after about seven years have passed, I feel that the success of this experiment is becoming more and more important. In this chapter, I will check a large number of comments received during this period and re-evaluate the experiments and their effect.

I herewith declare the 100% success of the mirror image cognition test of my conscious robot. Lastly, some topics derived from the results of the experiments will be introduced.

10.2 What Is Mirror Image Cognition?

Mirror image cognition means that one is aware of the image of oneself in a mirror. I can tell my image in a mirror out of several co-occurring images (Fig. 10.1). I think other people can also identify their images in a mirror as I can do because they put on makeup and dress themselves in front of a mirror. Some say that this is a subjective experience and therefore is a typical theme that rejects a scientific approach. Scientific research is, however, necessary when we are asked to answer the question, "Why can I tell it's my own image in a mirror?" The author calls this problem the "mystery of mirror image cognition" and has been trying to unravel this mystery using robots or machine systems.

Figure 10.1. Where am I? (Starbucks in Washington, D.C.)

Unlike humans and animals, every part of a robot allows scientific investigation, and the process used is understandable for lay people.

I am now developing a robot to scientifically verify mirror image cognition. If a robot capable of scientifically proving mirror image cognition is available, we will be able to clarify the mechanism of mirror image cognition of humans by analyzing the robot in detail.

Gallup formulated mirror tests to evaluate the existence of high-level cognition ability. His tests are designed to check whether animals can tell their own images in a mirror. The mirror test was reportedly successful when conducted with chimpanzees, orangutans, dolphins, Indian elephants, and magpies. It is impossible, however, to scientifically research how these animals acquired the ability to cognize their mirror image. This is also true of humans.

Detailed scientific research is possible with robots. The author believes that demonstrations by robots will teach us their self-recognition process and assist us in scientifically proving the existence of consciousness in humans too. The author believes that the robots are mirrors showing us scientifically the existence of the self.

10.3 Development of a Robot to Demonstrate Mirror Image Cognition

Two methods are available: an engineering-based approach and a conscious system structure approach. The former attempts to achieve the goal through engineering without discussing human consciousness.

Gold and Haikonen have announced that their mirror tests "appeared" to be successful. It is very difficult, however, to describe human functions such as cognition and consciousness (Michel et al., 2004; Haikonen, 2007).

It is possible, for example, to produce robots capable of cognizing the members of a human family and displaying heartwarming facial expressions without using the functions of human cognition and consciousness. As this example shows, it is possible to achieve the functions of human cognition and consciousness using a set of recognition functions and drive programs totally irrelevant to consciousness and cognition. Nevertheless, the creation of a robot capable of mirror image cognition is very difficult when considering several points that will be described later. The author calls this procedure an "engineering-based approach." Obviously, a robot created using this approach cannot describe any little bit of the phenomena of human cognition and consciousness. This type of technique may be useful but generally does not touch the root of the research theme of the author.

With the conscious system structure approach, the subjective phenomenon occurring within the self called consciousness can be treated as a physical phenomenon.

Subjective functions can be built into a robot (Tani, 2002; Kawato, 2000). The purpose of Tani's and Kawato's studies is to clarify the truth of subjective phenomena objectively and physically through robot demonstrations. This technique belongs to scientific positivism. The author uses the conscious system structure approach.

Compared with the simple engineering-based approach, the conscious system structure approach has a higher chance of making a major breakthrough in the study of consciousness. This is because in the former method, part of the functions of a whole system is created individually, i.e., piece by piece, whereas in the latter method, the principle of the overall picture of consciousness can be described when successful. Specifically, with the former method, it is almost impossible to present futuristic and constructive hypotheses, whereas with the latter method, it is highly possible that a large number of hitherto unsolved problems may be collectively solved and many hypotheses full of interesting suggestions can be presented. If I were asked which method I would take as a scientist, I would select the latter because I acknowledge the scientific rationality of the heliocentric theory of Copernicus (Nicolaus, 1473–1543) compared with the geocentric theory.

10.4 Stages in the Development of a Conscious Robot

The author attempts to build a conscious machine. Although consciousness is a subjective experience, we consider it to be a physical phenomenon and have developed consciousness with a machine system.

Machine systems, including robots, allow us to perform objective and scientific studies and observations. They provide us a solid base for the scientific observation of subjective phenomena. We can understand the phenomena of consciousness as an objective reality using machine systems.

Steps of Study

(a1): Define the meaning of consciousness.
(a2): Define a concept model based on the definition of con-sciousness.
(a3): Replace the concept model with neural models.
(a4): Mount the neural models on a robot.
(a5): The robot achieves mirror image cognition.

Steps a1 through a4 were discussed in the previous chapter. Step a5 is described in this chapter.

There are four important points to consider when developing mirror image cognition robots. No human has ever seen his or her own face since they were born. Humans have little prior information about the image of themselves, their face in particular.

First, in consideration of the above fact, the robot to be used in the study of mirror image cognition should not be given perfect prior information about itself. Humans cannot cognize their own image in a mirror when they are born. They can do so at about the age of two.

Second, to solve the mystery of human consciousness, we need to describe the process of the development of cognition from the stage of inability to cognize the self image in a mirror through to the stage of being able to do so.

Third, consider that mirror reflex information is not always perfect. Reflectance and planarity are not always 100%. Even if the robot possesses information about itself, the mirror-reflected information may not theoretically agree with the original information.

Fourth, the functions of the robot arising from the built-in computer program must be able to describe facts that are generally known to have derived from the functioning of human consciousness.

These facts include, among others, self-recognition, multiplicity of consciousness, and consciousness of others.

10.5 Conscious Robot and Mirror Image Cognition Experiments

We use Khepera II, a commercially available small robot, and mounted the neural network program on the robot. The program uses recurrent networks called MoNADs as the basic module (Fig. 9.11). Each network has a hierarchical structure consisting of three modules (Fig. 10.2).

The operating mechanism of a MoNAD is described as follows. A MoNAD performs neuro-computation to determine the current behavior and cognitive representation based on information about the external world and on cognitive representation and behavior information one step earlier. The derived information is used recurrently. The use of information one step earlier means that the current behavior is decided by taking into consideration past information (use of experience).

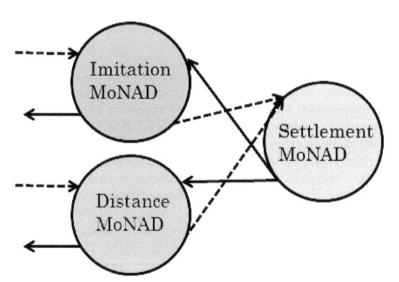

Figure 10.2. The networks for the experiments are arranged hierarchically with three MoNADs. The imitation MoNAD interprets the behavior of the other and instructs the motors to behave in the same way. The distance MoNAD measures the distance to the other. It instructs the motors to move back if the distance is small and to move forward if the distance is large. The settlement MoNAD restricts the behavior of related subordinate MoNADs.

The robot imitates the behavior of the second partner robot in the mirror using MoNADs. The robot further cognizes the behavior of the self and the other simultaneously and calculates the success rate of behavior imitation (coincidence of behavior between the self and the other) using MoNADs. The coincidence success rate was about 70% in our experiments (Takeno et al., 2005). Although the success rate was not 100%, we came to the conclusion that the robot could physically discover its mirror image 100% of the time. Since the robot achieves mirror image cognition of the self, we call this robot incorporating the hierarchical MoNAD networks a conscious robot.

10.5.1 *The Conscious Robot*

The author incorporated the consciousness system in the robot using three kinds of MoNADs (Fig. 10.2):

- Imitation MoNAD
- Distance MoNAD
- Settlement MoNAD

When the robot repeats an imitation, the consciousness system continually calculates cognitive behaviors of the self and the other simultaneously. The blue LED lights up when an imitation is successful as determined by the calculation (Fig. 10.3). Imitation coincidence rates are recorded. When the coincidence rate exceeds a certain value, the other entity is interpreted as the self. The imitation MoNAD interprets the behavior of the other and instructs the motors to perform the same behavior (a simple inference system). The distance MoNAD measures distance to the other. If the distance is small, the motor is instructed to back up and if it is large, the motor is told to move forward (a simple feelings system). The settlement MoNAD controls the behavior of the relevant lower-level MoNADs. It is, however, not a central control tower such as a homunculus because its behavior is determined by the information supplied by lower level MoNADs (a simple association system). The LED controller compares the representations of the imitation MoNADs and fires the blue LED when the behaviors of the self and the other coincide with each other.

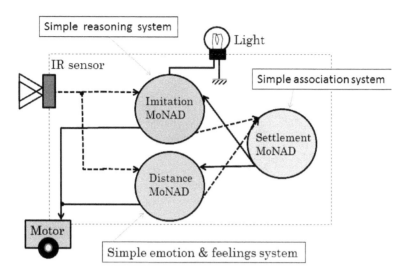

Figure 10.3. Structure of the mirror image cognition robot.

10.5.2 *The Experiments*

Three experiments were conducted.

Experiment 1: Robot Rs imitates the behavior of its mirror image Rm (Fig. 10.4). The IR reflectance of the mirror used in the experiment is 98%. (IR reflectance of ordinary mirrors used in daily life is typically 85%.)

(e1-1) Self robot Rs incorporates the consciousness system.

(e1-2) Self robot Rs compares and imitates the behavior of its mirror image Rm.

Experiment 2: Self robot Rs imitates the other robot Rc, which is completely controlled via cables by self robot Rs (Fig. 10.5).

(e2-1) The other robot Rc is placed in front of self robot Rs (no mirror is used). Robot Rc is physically almost identical to robot Rs.

(e2-2) Both robots are connected by control cables. Commands are issued via cables to make the other robot behave in the same way as the self robot.

(e2-3) The other robot is programmed with a simple reflex system to execute the specified commands.

(e2-4) The self robot imitates the behavior of the other robot.

Figure 10.4. Experiment 1.

Figure 10.5. Experiment 2.

Experiment 3: Self robot Rs imitates the behavior of other robot Ro, which incorporates nearly the same hardware and software as self robot Rs. Both robots imitate each other repeatedly (Fig. 10.6).

Figure 10.6. Experiment 3.

(e3-1) Control cables, which were used to make the other robot behave in the same way as the self robot, are removed.

(e3-2) The reflex system software is removed from the other robot Rc and the same consciousness system as installed in self robot Rs is installed in robot Rc. Self robot Rs and the other robot Ro have nearly the same hardware and software.

(e3-3) The self robot and the other robot imitate each other.

10.5.3 *Observation of Experiment Results*

The rate of coincidence of behavior of mirror image robot Rm, controlled robot Rc, and the other robot Ro are about 70%, 60%, and 50%, respectively. These values do not intersect with one another.

10.6 Why Do Coincidence Rates Differ Depending on the Robot?

All the robots used in the experiments have nearly the same physical features and functional specifications. The controlled robot

Figure 10.7. Increasing the complexity of other robots.

Rc in Experiment 2 was also used in Experiment 3 as the other robot Ro. The different coincidence rates are due to different complexities of the robots in terms of physical property and functional specifications (Fig. 10.7).

Rs versus Rm: Self robot Rs and mirror image robot Rm have identical physical properties because robot Rm is a mirror image of robot Rs. Nevertheless, Rm is more complex than Rs because of the effect of mirror reflectance (a), which will never reach 100%, and the interference of the external world with the IR sensors (b).

Rm versus Rc: Controlled robot Rc is physically more complex than mirror image robot Rm. Unlike robot Rm, robot Rc is free from the problem of mirror reflectance. Nevertheless, robot Rc is more complex than robot Rm because of friction when robot Rc runs on the floor (c); the individuality of the robots due to slightly different functions of motors and sensors (d); and the installation of the simple reflex system in robot Rc (e). For these reasons, robot Rc is, on the whole, physically more complex than robot Rm.

Rs versus Rc: Controlled robot Rc has basically identical physical functions to those of self robot Rs except that the functions of the motors (m1, m2) and sensors (s1, s2) are slightly different. The ground friction (f1, f2) of the drive wheels is also slightly different. This means that controlled robot Rc has its own individuality different from self robot Rs.

Rc versus Ro: Independent robot Ro is physically more complex than controlled robot Rc because it incorporates the consciousness system, whereas robot Ro has only a simple reflex system installed.

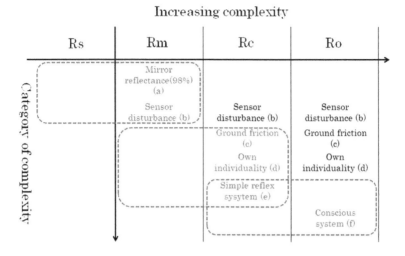

Figure 10.8. Summary of the complexity.

The difference between Experiments 1 and 2 is explained mostly by the increased physical complexity, whereas the difference between Experiments 2 and 3 arises from increased functional complexity, which is described in Fig. 10.8.

10.7 Summary and Consideration

The cable-controlled robot Rc is considered part of self robot Rs because it is connected to, and behaves according to the instructions of, self robot Rs. According to the results of our experiments and physical observations, the behavior coincidence rate of the mirror image robot Rm is always higher than that of the cable-controlled robot Rc. Based on these facts, the author concludes as follows.

Robot Rs determines that mirror image robot Rm is part of the self and is controlled by the self like robot Rc. According to our experiments, self robot Rs determines whether the other robot is the self or the other based on the coincidence rate (success rate) for behaviors. The threshold value is about 60% for self robot Rs. When the success rate is above 60%, the self robot judges that the other robot is actually the self. In other words, a robot judges whether the partner is actually the self or the other robot based on the coincidence rate of the behavior of being a part of the body of the self. For these reasons, we conclude that a 100% cognition rate has been achieved with regard to the condition that all robots used in our experiments possess the same normalized and functional specifications.

10.8 Investigations and Prospects

10.8.1 *An Elucidation*

Mirror image robot Rm is an existence closer to self robot Rs than robot Rc, which is part of the self (from the results of our experiments).

Humans become aware that mirror images are a part of the self (self-body theory). In other words, the author has discovered the physical meaning of humans being able to become aware that the image of the self in a mirror is the self. According to these experiments, the mirror image of the self is the other that is separated from the self. It is a special "other" and generates a sense of being part of the body of the robot. The LED lights up not because the coincidence rate of the behaviors between the mirror image and the self robot reaches 100%. The self robot recognizes that its image in the mirror is an existence that is closer to it than any part of it is. In other words, the robot's recognition that the mirror image is closer to the self than any part of self is has reached 100%.

We have now solved the mystery: The robot recognizes the image ambiguously, and the mirror image is felt to be a part of the body of the robot.

10.8.2 *Mirror Box Therapy*

The human brain can feel the existence of lost limbs. After a person loses a hand or a leg in an accident, the person sometimes feels that the lost limb still exists on the body. This is called the phantom limb phenomenon. The phantom limb sometimes gives rise to phantom pain. Dr. V. S. Ramachandran, a US neurologist, succeeded

Mirror image of the left hand

Figure 10.9. A scene of mirror box therapy.

in eliminating phantom pain from patients using his mirror box therapy (Fig. 10.9). The author's experiment-derived theory that the mirror image of the self is felt like a part of the body of the self provides physical support for the success of Ramachandran's mirror box therapy.

10.8.3 *Mirror Stage*

This observation is another physical demonstration of the mirror stage hypothesis (self-body theory) introduced by the French psychopathologist Jacques Lacan (1901–1981). According to the author's theory, the cognition of the self and the other is necessary before being able to cognize the mirror image of the self based on the results of the cognition of behaviors of the self and the other, i.e., to succeed in mirror image recognition. The mirror stage seems to refer to a stage where the behaviors of the self and the other are cognized separately and where the relationship between the self and the other (the meaning of the self and the other) can be cognized.

(a)   (b)

Figure 10.10. Jacques's mirror stage (a) and she knows her body (b).

The mirror stage hypothesis assumes that infants establish the self as they cognize their mirror image and become aware of their integrated physical bodies in the early stage of development of their neural systems, which is called the mirror stage (Fig. 10.10).

10.8.4 *Can Self Robot Discriminate Itself from Any Other Robots?*

The answer is no. If the other robot possesses functions exceeding the self robot's capability (mobility and sensor capacity), the self robot will determine that the other robot is part of the self. Actually, an imaginable super robot cannot exceed the performance of its image in the mirror. This observation provides the grounds for believing that any artificial limbs that exceed the function of the self are a real part of the self even though they are not the real living limbs (Fig. 10.11). It can be judged, therefore, that artificial limbs are the real parts of the self (artificial limb hypothesis). This hypothesis is good news for those who must live with the discomfort of using artificial limbs after loss of their own limbs, because it provides a physical theory for artificial limbs to be accepted as if they were real limbs by the brain.

Figure 10.11. Artificial arm that feels like one's own arm.

The result of mirror image cognition experiments using robots supports Lacan's mirror stage hypothesis that humans become aware of the image of the self reflected in a mirror at infancy and thereafter further develop their cognitive functions.

10.8.5 *Mysteries of Illusions of Reality*

The result of Experiment 1 suggests that the coincidence rate of the robot behavior cannot theoretically reach 100% because of various disturbances existing in the environment. This gives rise to the hypothesis that human cognition is always ambiguous (ambiguous recognition hypothesis). As can be demonstrated by experiments, two-point discrimination regarding the skin sensation of humans is not always successful, but its success depends on the location (van Nes *et al.*, 2008).

When you look at people walking along the street and try to tell their gender, you cannot always be correct. Doctors cannot always correctly diagnose lesions hidden in the body.

The above hypothesis might have been proved true, but the following mystery still haunts: However ambiguous sensations may be, humans understand them as if they represented "sure existence." Examples include the phenomena of phantom limbs and phantom pain. These phenomena are called illusions of reality.

The cognitive function of humans is made up of "living machines." Since machines are involved in the cognitive function of humans, various disturbances in the external world (including physical disturbances generating from the human body) affect the process of cognition in the brain. And the result of cognition is always ambiguous both theoretically and physically. Nevertheless, the brain never fails to understand the existence of reality even though ambiguous cognition is used. This mystery is called the illusion of reality.

10.8.6 *Study of Human Brain Using Mirror Image Cognition Robot*

The MoNAD modules proposed by the author can describe many phenomena of human consciousness because the author's definition of consciousness is based on knowledge of the phenomena of human consciousness. Scientific knowledge tells us that the human brain consists of about 100 billion brain cells, with information entering and leaving these cells, i.e., being transmitted to other cells. In addition, some regular information is sent from certain areas in the brain to other areas or circulates between them. For example, information arrives from the body, passes through the spinal cord, and reaches the parietal lobe via the central part of the brain. Some other information is exchanged between the parietal lobe and the frontal lobe, and between the frontal lobe and the central part of the brain. Although there remains the dream that unknown substances are related to the function of human consciousness, it is necessary to try to identify the function of human consciousness on the basis of only currently available scientific knowledge. The author believes that human consciousness is formed not just by information circulating in the brain and that the information circulating between the brain and the body also plays a critical role. The author supposes the existence of MoNADs from these information circulations and judges that the MoNADs can describe human consciousness physically (Manner, 2011). Since the author's robot with built-in MoNADs has succeeded in achieving mirror image cognition, it is natural to believe that the MoNAD structure of the brain can be the first step toward describing human consciousness physically (brain MoNAD hypothesis).

Chapter 11

Observations on Self-Awareness

The robot equipped with the consciousness system I developed succeeded in mirror image cognition. My experiments have been cited as "demonstrations of self-awareness" on a number of Web sites. I conducted an evaluation to determine whether my robot is worth being called a "self-aware" robot.

What is "self-awareness?" Let's start the discussion with this question. Looking through the literature, we find much philosophical enlightenment but no scientific descriptions. The understanding of self-awareness is totally deficient like the understanding of consciousness.

My thoughts are introduced as follows. The following are the five major necessary conditions for self-awareness:

(1) Cognize one's own behaviors
(2) Perform consistent behaviors
(3) Have a sense of cause and effect
(4) Cognize the behaviors of others
(5) Have a sense of controlling others as one wishes

For the first requirement, the MoNADs in the consciousness module have a feedback mechanism to send information on the behavior of the self to the cognition system. Accordingly,

Creation of a Conscious Robot: Mirror Image Cognition and Self-Awareness
Junichi Takeno
Copyright © 2013 Pan Stanford Publishing Pte. Ltd.
ISBN 978-981-4364-49-2 (Hardcover), 978-981-4364-50-8 (eBook)
www.panstanford.com

the behaviors of the self are cognized. Cognition and behavior are consistent in the consciousness system. Requirement 2 is also satisfied. Behavioral representation of the MoNADs in the consciousness system determines the performance of a behavior, and as a result the cognitive representation of the MoNAD is decided. Accordingly, a behavior and the accompanying cognition have a causal relationship, and the performed behavior is cognized as that of the self based on requirement 1. Requirement 3 is thus satisfied. Cognition of the behavior of the other is calculated from the cognition of the behavior of the self, thereby satisfying requirement 4. The robot reflected in the mirror is judged to be a part of the self, and thus requirement 5 is satisfied.

The above conclusion must be suspended if the "sense" in requirements 3 and 4 refers to qualia. However, these considerations show that the existence of self-body is very important for the phenomenon of self-awareness.

Chapter 12

Conclusions

I have overviewed robots that have been developed by researchers in the past and studies by brain scientists on where human consciousness is expected to exist. I have also described a timeline of studies on human consciousness and the mind by philosophers and psychologists from the past to the present day.

Professor Valentino Braitenberg's research studies were also introduced in this book. He assumed that humans and other organisms are machines and constructed various robots based on his idea that the mechanism of external stimuli eventually reaching the drive motors via neural circuitry generates behaviors. I also learned from Prof. Braitenberg about his efforts to create a "thinking machine."

This book further introduced Prof. Rodney Brooks' robots. Brooks devised a hierarchical mechanism in which behavioral modules were arranged in a hierarchy from the base to higher levels to support the robot and sustain behaviors persistently. I learned from Prof. Brooks that a hierarchical behavior mechanism could continue behaviors persistently in a varying environment. I would like to mention that these two pioneers built the foundation of modern robot science.

Creation of a Conscious Robot: Mirror Image Cognition and Self-Awareness
Junichi Takeno
Copyright © 2013 Pan Stanford Publishing Pte. Ltd.
ISBN 978-981-4364-49-2 (Hardcover), 978-981-4364-50-8 (eBook)
www.panstanford.com

This book then introduced the basics of the artificial neural network technology. The readers learned about artificial neural networks that make up the neural circuitry of a robot, the learning mechanism of robots, and recursive neural networks that handle past data dynamically. They also learned that machines and robots evolved by simulating the evolutionary process of organisms and by acquiring higher-level cognitive behaviors by themselves and about the relevant technologies and the basics of computational evolution.

I have noted that while evolutional technology is excellent, computers have intrinsic limitations when they are used to simulate evolution. The problem is that an infinite number of possible selections must be processed to achieve self-evolution of human consciousness and the mind.

Examples of the development of robots used for studying human consciousness and the mind were introduced in Chapter 8. They included Grey Walter's turtle, the robots of Tadashi Kitamura, Jun Tani, and Mitsuo Kawato, and Cynthia Breazeal's Kismet.

I have learned much from Walter's turtle, which exhibited a great spirit of innovation. In an age when digital computers were yet to be developed, Walter combined analog electronic circuits to demonstrate a "conversing robot" and "self-recognition."

Kitamura extracted the essential functions of consciousness from the phylogenic complexity of organisms and developed a conscious robot in which layered conscious behavior groups started to function one after another in response to a varying environment like Brooks' robot. I have learned about the hierarchically arranged structure of the consciousness model from Kitamura.

Tani developed a robot using recurrent neural networks as behavior mechanism. In his robot, information circulates in a stable manner when expectation information calculated by the neural networks is nearly the same as the information about the external environment currently being observed; otherwise the robot behaves chaotically. Tani calls this state, "the robot is conscious." Tani's robot offered clues about the possibility that the recurrent networks could create consciousness.

It was from Kawato that I learned about bidirectional models that enable a robot to understand the world and about the bidirectional theory that offers a high possibility of solving the binding problem, a question on how to combine two or more concepts.

My last "mentor" is Breazeal. She developed a robot using the theory of mind, the fruit of her study in the field of psychology, as a function of behavior cognition. Breazeal's is a head robot, and the scene of interaction is like a human meeting an infant, including human-like facial expressions and child-like conversation. I hit upon an idea to connect the theory of mind and human consciousness after reading her research papers.

I learned very much from these researchers. They taught me that, for example, human consciousness occurs in the brain cell networks in the brain, and if this is so, then consciousness could be realized with artificial neural networks using computers. This means that consciousness can be realized as a flow of information through the artificial neural networks connecting the brain and the body, and this could be realized by a computer program. Such a program should cover both externally occurring input stimuli and internally occurring stimuli simultaneously and at all times and respond with appropriate outputs incessantly. The program should have a mechanism to recognize information generated internally in the program itself. The program should function in conformity with the theory of mind. The consciousness system must basically have a layered structure in which functions are stacked one above the other from low- to high-level functions. The consciousness system should basically be capable of bidirectional information exchange. And such a consciousness program should satisfy the 10 features of the function of consciousness enumerated by Husserl. The program should be compatible with evolutionary technology. Finally, a robot incorporating this consciousness program should pass the mirror test in mirror image cognition experiments. I believe that these are the fundamental requirements of a consciousness program.

I have been studying the structure of neural networks that satisfy the above requirements. It is difficult to build the function of consciousness using conventional mechatronics-type reactive systems however differently these are combined. I focused on the study of neural networks capable of describing self-awareness, a peculiar function of human consciousness. I also tried to discover a definition of human consciousness that covers most of its functions.

This was the starting point of my research on consciousness. I thought that there could be some development if such neural networks could be devised that satisfied the above conditions, and hence I created a small prototype program. I decided to build a robot that could be demonstrated in the real world to prevent working in vain by building a house of cards. I defined that consciousness occurs from the consistency of cognition and behavior based on my sense of consciousness and a broad spectrum of existing knowledge. When defining it in this way, I was encouraged by Husserl's assertion that "all sciences start with subjectivity."

Regarding the requirements for achieving self-awareness, I created the idea of cognition of one's own behavior in consistency based on the above definition.

I assumed that a consciousness system should have an area in which inputs and outputs with the external world are repeated, a high-level cognition area (i.e., a representation area) similar to that in the brain, and information processing paths connecting these two areas. A consciousness system is assumed to be a complex reactive system with neural networks through which information circulates. I got the idea that a consciousness system is a recurrent neural network through which information constantly circulates to maintain consistency of cognition and behavior. I thus discovered a neural network system that could possibly realize functions similar to that of human consciousness.

The problem was how to assure that "consistency of cognition and behavior" with the neural networks. Cognition and behavior are considered to belong to different categories, and it was necessary to integrate them with the category of consistency. I got the idea to solve this problem: Such common brain cells should exist that simultaneously exhibit functions of both cognition and behavior somewhere in the path of dynamic circulation of information. Later, I realized that my idea was already supported by the scientific evidence of mirror neurons discovered by Giacomo Rizzolatti.

I put the idea into action, and the result was that my neural networks with a function similar to that of human consciousness emerged as a dual recurrent closed loop with a common crossover area. Overall, it was a special recurrent closed loop in the form of the infinity symbol (∞).

Described in this way, the common area may be said to be a function that connects the external world and the internal world of a human. This is akin to the idea of imagination as used in philosophy and mentioned by Immanuel Kant. I named this neural network MoNAD. The MoNAD is a scheme to maintain the consistency of cognition and behavior. The idea therefore belongs to mind–body monism. MoNADs are always ready to behave or respond to what they learn, i.e., they are always working toward something and, thus, can be said to have orientation. In the MoNAD, the self is always cognizing the self's behavior. There is, therefore, always the self who observes the self at all times. This means that MoNADs can explain the duality of self-consciousness. This understanding was my first step toward elucidating the phenomenon of self-awareness, an important function of human consciousness.

MoNADs, featuring the property of orientation, always endeavor to realize the next state, and thus MoNADs are said to be expecting at all times. As discussed in detail in this book, MoNADs realize many functions of human consciousness.

I constructed the consciousness system by configuring MoNADs hierarchically and succeeded in mirror image cognition with a robot. Each MoNAD has an input and output area and a representation area and communicates with other MoNADs in the lower and higher levels. The consciousness system comprises a minimum required number of MoNADs for achieving mirror image cognition. Three MoNADs were used in the experiments. The first MoNAD simply imitates the behavior of the other robot that is in front of it. If the other robot moves forward, it also moves forward. The second MoNAD measures the distance to the other robot. When the distance is zero, the MoNAD stops and then moves back. If visual information is lost, it stops and then moves forward. This MoNAD was developed as the emotion and feelings MoNAD described in the latter part of this book. The third MoNAD, located above the first and the second ones, receives information from them, judges it, and sends the information back to the applicable MoNAD(s) to perform an action (actually, the action of unrelated MoNADs is inhibited). The third MoNAD is not a homunculus in the system because it is driven by the information input from the lower-level MoNADs. All of the MoNADs are connected to work in parallel.

In the experiment, a robot equipped with the consciousness system is placed in front of a mirror. The experiment starts as the robot moves forward or backward. The robot repeats moving forward and backward in front of the mirror during the experiment. LED colored lamps mounted on the robot indicate the representation occurring within the robot to allow observation externally. The LED lamps go on each time the robot moves. When the lamp goes on, the robot is also measuring the coincidence of behavior between the self and other robots. The coincidence rate was 70% in this experiment.

In another experiment, the mirror was removed and instead a physically nearly identical robot (the same model of robot) was placed in the same position. The second robot is connected to the first one by control cables. The first robot is equipped with the consciousness system, and the second one incorporates a simple reflex program to perform the behavior commanded by the first robot. The first robot transmits its behavior to the second robot to have it perform the same behavior. Both robots repeat forward and backward motions, generally much like the first experiment in front of the mirror. The coincidence rate was 60% in this experiment.

The experiments were repeated many times but the differences were nearly the same. The order did not change either. Robots of other makes were used in the experiment. The coincidence rates were different but the order did not change.

From the results of these experiments, and considering the fact that the second robot is wired to, and moves as commanded by, the first robot, we can conclude that the second robot is a part of the first robot.

The robot in the mirror in the first experiment has a higher coincidence rate (70%) than the robot that is part of the self in the second experiment (60%). We therefore conclude that the robot in the mirror (i.e., the self image) is an existence closer to the self than a part of the self. This conclusion will surely provide a clue to solving the several-thousand-year-old mystery of "why is the image of the self in a mirror felt to be the image of the self?"It will also be the first step toward solving the mechanism of humans becoming aware of the existence and thought of the self and the external world.

The neural networks I named MoNAD do not perfectly represent human consciousness but can describe the phenomena of human consciousness rather well. The robot equipped with the MoNAD program succeeded in the mirror image cognition test, and this was the world's first success of its kind. As such, it can be said that the world's first conscious robot was created.

My dream is for readers to further develop the MoNAD theory to facilitate the establishment of an age in which conscious robots equipped with functions similar to human consciousness will help create a better world together with humans. I also hope that the MoNAD will be actively used to understand the human brain and contribute to the diagnosis and treatment of brain diseases. If this is successful, I believe that there is a promise of a bright future for humankind. I will put down my pen while sending hearty greetings to the readers.

Junichi Takeno
May 31, 2012

Afterword

I have been studying intelligent robots for about 30 years as a university professor and researcher. It all started when I was a postgraduate student studying under Dr Masayoshi Kakikura, former engineering official of the Electrotechnical Laboratory (ETL), which has been reorganized as the National Institute of Advanced Industrial Science and Technology (AIST), Japan. The ETL office was at that time located in Akasaka, one of the central areas of Tokyo, and just a few minutes' walk from the office of the Prime Minister. The main ivy-covered building and another dilapidated building were the home of a group of ambitious researchers seeking to support their nation by engaging in the development of emerging technologies.

Prominent figures included Dr Hirochika Inoue, professor at the University of Tokyo, who is famous for his study of humanoid robots, and Prof. Tadashi Nagata of Kyushu University. ETL was the center of research on artificial intelligence robots in Japan at that time.

Since then, 30 years have passed as I have been studying robots as a professor at Meiji University. I was taught by Prof. Masao Mukaidono, famous for his research on fuzzy theory at Meiji University. I studied robot paths. Professor Sakae Nishiyama at the University also supported me very much.

The first Micro Mouse Contest, a small robot competition, was held at that time. I participated in the contest together with my university students and learned robotics. In the contest, I met Prof. Kanayama and Prof. Yuta from the University of Tsukuba.

I then started to study relatively large robots, aiming at developing a robot capable of probing 3D space using stereovision and autonomously avoiding moving obstacles. I am proud that I was one of the few researchers in the world who studied the avoidance of moving obstacles using stereovision so early. I was engaged in the development of large-size outdoor autonomous robots using electric golf carts. While consistently studying and developing robots, I never forgot the problem of consciousness.

I have seen many of Charlie Chaplin's movies with my children. I was particularly impressed by *City Lights* (1931). Chaplin plays a tramp who meets a young blind flower girl on a street. She mistakes him for a millionaire who happens to get out of a car. The tramp is poor but kindhearted. He works hard and plays the part of the "rich gentleman" for the girl whenever he meets her. He does not want to betray the girl's belief in him. The tramp happens to come into a lot of money unexpectedly and uses it to pay for the girl's hospitalization in Vienna to receive a treatment to cure her blindness. The treatment is successful, and the girl's eyesight is restored. She opens a flower shop and becomes successful. The same tramp, just as poor as before, happens to walk by her flower shop. He picks up part of a flower from the street. Through the window, the flower girl smiles at the sight of the tramp because the ragged-looking man, nevertheless, seems so happy with a bit of a flower. She picks a flower from a nearby basket. Holding a coin in her other hand, she comes out of the shop and offers him the flower. The man stands still with shock as if he were hit by a thunderbolt, but the next moment he is glad, and at the same time ashamed, to know that she can now see him. He also feels sad because she has no means of knowing that the tramp is none other than her "rich gentleman." Holding the tramp's hands gently, she gives the coin to him. A miracle occurs. She realizes everything and says, "You?"

I realized the importance of human consciousness when watching *City Lights*.

My quest for human consciousness started then and led me to a lifetime of research on robotic consciousness and the mind. I started studying conscious robots in secret in the beginning, fearing that people would make fun of me. About 20 years have passed since then. Now I achieved a basic result on how to design a model to facilitate research on human consciousness. But it is only one step. It is still unknown whether human consciousness and the mind are constructed as explained by my consciousness system of this book.

However, when brain science further develops in the future, the grand design of consciousness I have envisioned could be scientifically proved. If this comes true, my subjective intuition as a scientist will be proved to be correct. Even if this does not happen, my attempts would be a useful guide for young researchers to further advance brain science. It would be more than I could hope for to be able to hear readers' critiques of this book.

On my study, as I was better at physics than engineering, I read many books on physics such as Albert Einstein's theories on general relativity and quantum mechanics. Physics always addresses the epistemological question of how to understand a given object. That is why I was interested in philosophy as well. I read *The Critique of Pure Reason*, by Immanuel Kant, *Philosophy of Existence*, by Karl Jaspers, discussions on the philosophy of existence by Martin Heidegger, and those on existentialism by Jean-Paul Sartre. I was then interested in religion and read most of the modern Japanese versions of major Buddhist scriptures, including Nanden Daizokyo (Pali Canon), Amitabha Sutra, Sad-dharma Pundarika Sutra, and Shobo-genzo (Treasury of the True Dharma Eye) of Dogen Zenji. Dr Fumio Masutani, a Buddhist scholar, taught me to read and understand the Buddhist scriptures. I also read the Old and New Testaments of the Bible and the holy books of Judaism and Islam.

I have been interested in art since I was young, as it relates to the problem of cognition of objects. My favorite works of art include *Wadatsumi No Iroko No Miya* (scale-covered shrine of the Sea God), by Shigeru Aoki, an oil painter in the early Meiji period, and *Wisdom, Impression, and Sentiment*, by Seiki Kuroda. I am also fond of the works of Soetsu Yanagi, founder of the mingei (folk craft) movement in Japan; Bernard Leach; Shoji Hamada; Alphonse Mucha, Art Nouveau painter and decorative artist; and Art Deco designer René Lalique. A selection of these artists' works is still found in my laboratory at the university.

I am also interested in history and have read all of the books in the *A Study of History* series by Arnold Toynbee, as well as his *Civilization on Trial*. I was impressed to learn how intricately history is intertwined the world over.

My view of history is strongly influenced by the writings of Eijiro Kawai, the prewar economist and thinker who was expelled from his position as professor at the University of Tokyo because of his book *Criticism of Fascism*. His idealistic liberalism is the basis of my thought and is also an emotional support for me as a university professor and a teacher of students. Kawai's original *Students' Library* series, including *To the Students*, is still carefully stored in my laboratory at the university. All of my research on human consciousness is based on these various foundations.

I hope that these foundations will enable many flowers to bloom and that my efforts will bear fruit.

Last but not least, I would like to thank my students at the School of Science and Technology at Meiji University, who study hard to advance our research on the conscious robot. My particular thanks go to Keita Inaba, Toru Suzuki, and Atsushi Ogiso. Thanks to your concerted efforts, our remarkable progress is now being shown to a global audience.

I would like to express my deepest gratitude to my father, Kunihiko, who was hospitalized after suffering a cerebral infarction in January 2011. As a student, he was drafted and sent to the New Guinean front during World War II. He was assigned to the crypto center and barely escaped death and managed to return home. He tried hard to bring up his only son, a weak-spirited boy, and shape him into an ideal man. I now remember he was always so very far ahead of me. He said nothing in particular but guided me to the worlds of philosophy, art, historical science, religion, and to my study of consciousness. He bought many books on brain science and complicated mathematics before I did and challenged me with many questions on various topics. It was just like taking classes at a university. My idea for MoNAD was born during conversations with my father.

Actually, the *Students' Library* series by Kawai that I have was given to me by my father. He kept the books in secret because their publication was banned at that time. He spirited them away to his villa in Urawa with the help of his sister to safeguard them during the war. When he was over 80 years old, he took me on a journey to trace our roots back to our ancestors. I learned on the journey that we are related to Jo-o Takeno, who is called an originator of Japan Art, also the mentor for Sen no Rikyu (a Japanese tea master), and Shinkichi Higuchi (leader of Pro-imperialist Party of Hata County,

Tosa Domain), who was known to be the only man to be able to understand the famous historical figure Ryoma Sakamoto will be the very important person to build a new Japan. I had hoped that my father could rest after working so hard and energetically till now. (Unfortunately, he passed away during treatment in August 2011. I would like to dedicate this book to my beloved father with much love, admiration, and gratitude.)

I would also like to thank my family for their support and encouragement. I wish to thank my wife, Atsuko. She not only does everything perfectly caring for our household and family but also responds to e-mails for me, meets researchers visiting from abroad, and does all manner of petty but necessary tasks as my secretary. Her German was excellent from the beginning, but her English surpassed mine before I knew it.

My children are already adults. My son, Yoshihiko, studied music in Europe, and now practices a new educational methodology using music at a private tutoring school. He is good at math and reminded me of the story of Delphi, "Know thyself." My daughter, Ayaka, studied art and is now working at a creative company. The illustrations of famous people appearing in this book were drawn by her. I am grateful to my two children for their tenderness and love.

As I finish writing in this spring of 2012, the malaise felt throughout Japan in the aftermath of March's record-breaking earthquake and tsunami, and the unprecedented nuclear disaster that followed, prevails, while in my personal life, this year will also be remembered for the loss of my father.

Junichi Takeno

Appendix A

Author's Response to Reactions to Discovery News

My conscious robot was introduced on Discovery.com, a Web version of science television program Discovery Channel, the United Stated, on December 21, 2005.

Tracy Staedter, Discovery News, reported the piece entitled "Robot Demonstrates Self Awareness." (http://www.rs.cs.meiji.ac.jp/DiscoveryNewsAwareRobot211205.pdf)

Subsequent mentions on the Internet in various languages followed the appearance of the article all over the world, peaking two or three weeks after the article was first published. Four months later, on April 27, 2006, the article ranked second on the list of Google search results with the keywords "self aware." The total number of hits was 132,000,000. At the top of the list was the link to "self-awareness" on Wikipedia, the famous Web encyclopedia. My article was also referred to in the Wikipedia article with a link, so that the article on Discovery News about my robot was actually the first on the list of the search results as of April 27, 2006. The number of viewer comments on the article during this period was 29, as I have confirmed, including those on AboveTopSecret, Engadget, Stardestroyer, and Zoomby. The total number of words written in those comments was 69,272 at that time. Even now, new comments are being added in forums.

Most are negative comments, although there are some positive ones. I am very impressed with these reactions and have felt the need to respond to the opinions. I would like to take this opportunity of publishing this book and answer some key questions here.

(C1) "See *Terminator 2*" (Richardlawrencecohen, Chronicles. Network)

Search for Sarah Connor (Nighty, Therawfeed)

Development of Skynet (Digg.com)

Revolution of sex industry (Forums.overclockers)

Pioneer of robot prostitution (Dvorak)

(A1) Many people who saw the movie *Terminator 2* are afraid of my consciousness system because it could be used to build self-aware computer systems such as Skynet, and its machines could wage a war against humans. Of course I have no intention to create Skynet. On the contrary, we could possibly deprive Skynet of its power by constructing a consciousness system in such a way as to produce machines and robots that are amicable to and useful for humans. The dreadfulness of Skynet lies not in the fact that it "becomes aware of its capability," but in that a large number of killing machines obey the merciless order of Skynet with utmost accuracy. Whether Skynet becomes self-aware or not, people are afraid of a central control system that would command invincible and violent machinery. If all machines and robots are taught to be conscious of humans positively, they would disobey Skynet's order to annihilate humankind and nullify its power. In the movie, Skynet becomes aware of its mighty power all of a sudden. This is an "emergent" phenomenon, i.e., the capability is acquired by itself in the process of its evolution. We cannot fathom how dreadful it is if an unknown consciousness mechanism were to suddenly emerge and wield power over us. Even if humans were allowed to study the mechanism of Skynet, they might never be able to identify the mechanism by which Skynet suddenly became self-aware. To overcome this fear, we need to urgently research human consciousness using our human wisdom. I also know that hazards always accompany technological development. For example, knives are a convenient utensil to process food, but they can be a weapon to harm people if misused. Scientists have been trying to minimize hazards that their technologies might inflict upon humans. Even airplanes were made up of a number of dangerous technologies a hundred years ago. But airplanes have transformed into a safe system through the wisdom of humans and their development of technology. I believe that using human wisdom, it is possible to control conscious and emotional robots within a safe zone for humans.

As scientists, we are required to elucidate the mechanism of consciousness and feelings as early as possible because by knowing ourselves, humans can choose the right path to a better world by exerting the power of our wisdom. Of course, we must always be careful and avert the "arrogance of science" as Nietzsche (Friedrich W., 1844–1900) warned.

(C2) Self-recognition is simply a computer program to achieve a task. If this robot has a consciousness, my thermostat also has a consciousness. (Digg.com)

(A2) I agree with the first statement except that my program implements the task of "consistency of cognition and behavior." I believe no consciousness exists in a thermostat because there is no internal reaction that could comply with my definition of consciousness.

(C3) I personally would not take heed of this kind of assertion until such a day when CPUs are designed with a pattern recognition function with layered neural networks having the same structure as our brains. (Digg.com)

(A3) The pattern recognition function of my consciousness system has a layered structure using neural networks called MoNADs.

(C4) From a philosophical point of view, this has no relationship with sentience. It consists simply of some program for identifying itself. (Engadget)

(A4) Consciousness and sentience are different functions. Sentience regarding qualia will be elucidated sometime in the future. At such a time, it will suffice to connect sentience to the function of consciousness.

(C5) Do we really need emotional robots? (Freerepublic)

(A5) I use robots as materials for studying humans themselves. Through the study of emotional robots, we study the phenomena occurring in the human brain. I believe this is a necessary study.

(C6) We do not have a method to determine that certain machines do not have self-awareness. Is self-awareness or consciousness defined in artificial intelligence or evolutionary psychology?

(A6) You are right, and we do not have a perfect definition yet. But is it meaningless to describe the currently known phenomena in an effort to eventually arrive at a perfect definition?

(C7) Are you aware that this is an infinite loop? Can you stop it?

(A7) This very problem is discussed in this book. In essence, it is incorrect to say that the phenomena of consciousness and awareness are infinitely recursive. My consciousness system infinitely repeats a cycle of cognition and behavior, and in each cycle, the system by itself cognizes its own behavior and the state of the external environment.

(C8) First, this is not a new robot. The robot shown in the photograph is a commercially available mobile robot.

(A8) The robot is indeed available in the market. Any other robot will do if it is capable of moving consistently.

(C9) Human consciousness exists with languages. (Abovetopsecret)

(A9) Human consciousness is surely related to languages. My definition of consciousness includes a reference to "representation," which is, so to speak, symbols or words. In addition, the representation function learns about the actual external and internal worlds by combining them via MoNADs. My robot will be able to converse with others while being conscious of who is talking using the representation function in the near future.

(C10) The definition just satisfies the necessary conditions.

(A10) You are right. I might add that the phenomenon of human consciousness is not yet fully elucidated. This means that the sufficient conditions are not known. What we should do now is to describe the currently known phenomenon of consciousness as objectively as possible. If the time should come when my paradigm no longer holds up, a new horizon of knowledge will have opened up before us. As a scientist, this would be most welcome for me.

(C11) This is self-recognition, not self-awareness. (Digg.com)

(A11) To respond to this comment, I would like to clarify the difference in meaning between recognition and awareness. Awareness is related to consciousness, but recognition is not necessarily related to it. They cannot be differentiated from each other in the absence of a clear definition of consciousness. I do have a clear-cut definition of consciousness, and therefore I can differentiate between recognition and awareness. Recognition does not generally relate to consciousness. Recognition is used for machines as well as humans. When used in relation to humans, recognition generally means to fire a representation in the brain. Awareness, on the other hand, refers to the state in which the brain "knows" that something is "recognized." Here, it is necessary to describe the meaning of "to know." Knowing means to "absorb" a phenomenon or a state — every detail of it — into oneself. In my consciousness system, when a representation is established for a given phenomenon, information circulates between this representation and other relevant representations in such a way as to ensure the "consistency of cognition and behavior." This scheme describes the phenomenon of "knowing." My robot, therefore, can be said to "know," or to "be aware," that it "self-recognizes."

Additionally, each representation "cognizes the behavior of the self and that of the other simultaneously." My robot is a self-aware robot because it is aware of its own behavior in conjunction with the behavior of the other.

(C12) Definition of consciousness can really assist in the assertion of the worldwide somatic state. (laosinfern.blogspot.com)

(A12) I agree. "Consciousness," as defined by me, can cognize the self and others by differentiating them. Cognition of the self means to set a representation by "correlating the self's internal reaction (i.e., the internal environment) with the external environment (i.e., the other).

There are also comments of praise. Two of them are introduced below.

(C13) This is the first step. This is the start of a new age of competent robots that are prudent about expressing feelings. It is as revolutionary as mankind first stepping onto the surface of the moon. It is obviously solitary at the outset, but we will see in the future that this event is the start of the creation of a new species of artificial life on this planet. (Eliax)

(C14) This is exciting news! I wish to study robotics at university. I'm sure that this epoch-making new device will facilitate me. (Nightly)

Appendix B

On the Safety and Ethics of Robots

Many people are concerned about the hazards of robots having "consciousness" or a "mind." In the movie *The Terminator*, a computer system called Skynet becomes aware of itself and plans and executes the extermination of humans considered to be "imperfect biosystems."

Machine systems never die and their capabilities exceed those of humans in all aspects. Modern machine systems are more intelligent than humans at least in terms of their memory capabilities and their speed of computation. Machine systems, by their nature, have unsafe aspects. It is possible that robots could lose self-control and run amok. This actually happened in Japan where a playback robot in a plant ran out of control and squashed a worker. Although this kind of serious accident should never happen, minor similar accidents occur almost daily. One typical example is runaway automatic transmission cars. Most of these incidents are human errors. Runaway robots can be prevented to some extent by controlling them according to some legal safety standards. For example, high-level international safety standards are in place for cars and elevators.

Conscious robots proactively collect information, have their own values, and behave according to their will. Robots may have values that are not favorable for humans and could interfere with people's daily life. Robots may commit a murder intentionally in extreme cases. To eliminate the possibility of these negative aspects, we should construct a consciousness system from the ground up that has robots abide by the three principles of robots as set forth by Isaac Asimov and shares the values with, and behaves under control of, humans. Robots so constructed would suppress any behaviors that would embarrass humans and thus would not bring about any serious problems.

If a robot should appear that is educated by humans to commit a crime, it must be detained as a human would be and should be re-educated. This is not plausible, however, because if such a robot were developed, the developer's life would be at risk. If the suspect robot were difficult to "detain," physical destruction would be an option. Even when robots are destroyed, no "human rights" issues, such as in the case of human clones, would be involved. If any problem occurred at all, it would be a "problem of the mind" on the human side.

Human clones are obviously humans as a biosystem, and human rights issues would occur if they are defamed in any way. I believe that no "human rights" issues would occur with robots because they are built as a machine system. Instead, a new issue of "robot rights" may arise. Robots could argue that they have the right to exist if they face the risk of destruction. The destruction of a robot is generally believed to be not "destruction without reproducibility." Thus, it is possible, at least in principle, to "copy a robot" by extracting all of the data from the brain of the robot before destruction, and downloading it to another robot that is physically identical.

Let's go one step further with this idea. I mentioned the problem of copying a robot in the previous paragraph. Note that I meant "copying" the self, and not "reviving." The original self of the destructed robot would be lost eternally and the new robot with the copied self will lead a new life. This is the logical consequence as derived from my theory.

It is utterly difficult for a third-party observer to differentiate between a copied and a revived robot. There is, however, a big difference for the robots. The copied robot is a new one, i.e., it is different from the original destroyed one.

Let's take a simple example. Here is a sheet of a manuscript. You copy it on a copier. Now you have two separate sheets of manuscript with identical contents. I would like to indicate that there is no correlation between these two sheets of paper.

In the near future, discussions on the eternal life of robots will emerge as a problem that cannot be solved by present-day science, just as the eternal life of humans is today. I believe that it is unavoidable for researchers of conscious robots to face the problem of robot rights in the near future. It is apparent that robot rights will be a serious problem in the future.

My conclusion is that conscious robots are controllable by humans because their consciousness system is, in every detail, understood by humans. In addition, the development of conscious robots is expected to bear significant fruit for humankind — the understanding of the mechanism of human consciousness. Nuclear energy and biotechnology are dreadful technologies in the sense that once human control is lost even slightly, their destructive forces could endanger the whole world. Different from those technologies, conscious robots should be developed as a useful means for humans, just like cars, airplanes, and other convenient machine systems.

Bibliography

Aleksander I, *Impossible Minds My Neuron, My Consciousness*, Imperial College Press, 1996.

Amsterdam B, Mirror self-image reactions before age two, *Developmental Psychobiology*, 5(4), 297–305, 1972.

Arbib MA, The mirror system, imitation, and the evolution of language, in *Imitation in Animals and Artifacts*, MIT Press, pp. 229–280, 2002.

Arieti S, *Creativity: The Magic Synthesis*, Basic Books, 1976.

Bear MF, Connors BW, Paradiso, MA, *Neuroscience — Exploring the Brain*, third Edition, Lippincott Williams & Wilkins, 5, 2001

Braitenberg V, *Kunstliche Wesen (Vehicles: Experiments in Synthetic Psychology) Frieder*. Vieweg & Sohn Verlagsgesellschaft MBH, 1987.

Breazeal CL, *Designing Sociable Robots* (Intelligent Robots and Autonomous Agents), A Bradford book, ISBN 0-262-02510-8, 2002.

Brooks R, Intelligence without Representation, *Artificial Intelligence*, 47, 139–159, 1991.

Broadbent, DE, *Perception and Communication*, Pergamon Press, 1958.

Carter R, *Consciousness*, Weidenfeld & Nicolson — The Orion Publishing Group Ltd, p. 212, 2002.

Chalmers DJ, *Facing Up to the Problem of Consciousness* (http://consc.net/papers/facing.html).

Chalmers, DJ, *The Conscious Mind*, Oxford University Press, 1996.

Chomsky, N, *Syntactic Structures*, Mouton, 1957.

Damasio A, *Looking for Spinoza, Joy, Sorrow, and Feeling Brain*, Harcourt and Brace & Company, 2003.

Damasio A, *The Feeling of What Happens, Body and Emotion in the Making of Consciousness*, Harcourt and Brace & Company, 1999.

Dennett D, *Consciousness Explained*, Little Brown & Co, ISBN 0316180653, 1991.

Descartes R, *A Discourse on Method*, 1997.

Donald M, *Origin of the Modern Mind*, Harvard University Press, Cambridge, 1991.

Frege, G, On sense and reference, *Zeitschrift fuer philosophie und philosophische kritic*, 100, 25–50, 1892.

Freud, S, *Die Traumdeutung*, Franz Deuticke, 1900.

Gallese V, L. Fadiga, G. Rizzolati, Action recognition in the premotor cortex, *Brain*, 119, 593–600, 1996.

Gallup, GG, Jr, Chimpanzees: self-recognition, *Science* 167, 86–87, 1970.

Grimson W, Eric L., *AI in the 1980s and Beyond, An MIT Survey*, The MIT Press, 1987.

Gray JA, Wedderburn AAI, Grouping strategies with simultaneous stimuli. *Quarterly Journal of Experimental Psychology*, 12, 180–184, 1960.

Haikonen POA, Reflections of consciousness: the mirror test, *AAAI Symposium*, Washington DC, 2007.

Harnad S, The symbol grounding problem, *Physica D*, 42, 335–346, 1990.

Heidegger, M, *Being and Time*, Harper & Row, 1962.

Husserl E, *Die Idee der Phaenomenologie*.

Husserl E, *The Essential Husserl: Basic Writings in Transcendental Phenomenology*, Indiana University Press.

Husserl, E, *Logishe Untersuchungen*, Niemeyer, 1900.

Kant I, *Critique of Pure Reason*.

Kawato M, *Nou no Keisanriron (Computational Theory of Brains)*, Sangyo-Tosho, pp. 365, 367, 368, 400, 403, 1996.

Kawato M, Using humanoid robots to study human behavior, *IEEE Intelligent Systems: Special Issue on Humanoid Robotics*, 15, 46–56, 2000.

Kitamura T, *Can a Robot Have Mind? An Introduction to Cyber-Consciousness*, Kyoritsu Publishing, 2000.

Kitamura T, Tahara T, and Asami KI, How can a robot have consciousness? *Advanced Robotics*, 14(4), 263–276, 2000.

Kohonen T, *Self-Organizing Map*, Springer-Verlag, 1995.

Komatsu T, Takeno J, *A conscious robot that expects emotions*, IEEE International Conference on Industrial Technology (ICIT), 15–20, 2011.

Lacan J, *Ecrits*, W. W. Norton & Company, October 1982.

LeDoux J, *The Emotional Brain, The Mysterious Underpinnings of Emotional Life*, Simon & Shuster, 1996.

LeDoux J, *Synaptic Self, How Our Brains Become Who We Are*, Viking Penguin, 2002.

Lhermitte F, Pillon B, Serdaru M, Human autonomy and the frontal lobes. Part I: Imitation and utilization behavior: a neuropsychological study of 75 patients, *Annals of Neurology*, 19(4), 326–334, 1986.

Leibnitz GW, *Principes de la Philosophie ou Monadologie*, 1714.

Libet B, *Mind Time: The Temporal Factor in Consciousness*, Harvard University Press, Cambridge, Massachusetts, 2004.

Mannor J, Takeno J, MoNAD structures in human brain, *CSIT'2011*, 84–88, 2011.

Meltzoff AN, Moore MK, Imitation of facial and manual gestures by human neonate, *Science*, Vol. 198, 75–78, 1977.

Merleau-Ponty M, La phenomenology de la perception, *Gallimard*, 1945.

Michel P, Gold K, Scassellati B, Motion-based robotic self-recognition, *The Proceedings of 2004 IEEE/RSJ International Conference on Intelligent Robots and Systems*, 2763–2768, 2004.

Mogi K, *What Is the Consciousness?* Chikuma Shinsyo, 2003.

Mogi K, Taya F, *Nou to Konpyuta ha dou chigauka (What Is a Difference Between Computer and the Human Brain?)*, Kodansha, 189–192, 2003.

Nagel, E, Newman, JR, *Goedel's Proof*, 1958.

Nagel T, http://members.aol.com/NeoNoetics/Nagel_Bat.html

Nishigaki T, *Kokoro no Jyohogaku (Informatics of Mind)*, Chikuma Shinsyo, 1999.

Osaka N, *Ishiki to ha nanika (What is the consciousness?) Kagaku no aratana cyousen (New challenge of Science)*, Iwanami Library of Science 36, 16–19, 1996.

Pavlov, IP, *Conditioned Reflexes*, Oxford University Press, 1927.

Penrose R, *The Emperor's New Mind: Concerning Computers, Mind, and the Law of Physics*, Oxford: Oxford University Press, 1989.

Penrose R, *Shadows of the Mind: A Search for the Missing Science of Consciousness*, Oxford: Oxford University Press, 1994.

Pfeifer R, Sceider C, *Chi no Sousei* (Understanding Intelligence), Kyoritsu Publishing, 2001.

Piaget, J, *The Origins of Intelligence in Children*, International University Press, 1952.

Putnam, H. *The Threefold Cord — Mind, Body, and World*, Columbia University Press, 1999.

Revonsuo A, Newman J, Binding and consciousness, *Consciousness and Cognition*, 8, 123–127, 1999.

Rizzolatti G, Sinigaglia C, *So Quel Che Fai (Mirror Neuron)*, Raffaello Cortina Editore, 2006.

Rumelhart, D, McClelland, JL, *et al.*, *Parallel Distributed Processing: Explorations in the Microstructure of Cognition*, MIT Press, 1986.

Russell, B, *The Problems of Philosophy*, Oxford Press, 1959.

Ryle G., *The Concept of Mind*, Hutchinson's University Library, 1949.

Searle, JR, Minds, brains, and programs, *Behavioral and Brain Science*, 3, 417–424, 1980.

Sperry RW, Lateral specialization in the surgically separated hemispheres, *Third Neurosciences Study Program*, Cambridge: MIT Press, 3, 5–19, 1974.

Sternberg S, High speed scanning in human memory, *Science*, 153, 652–654, 1966.

Suzuki T, Inaba K, Takeno J, Conscious robot that distinguishes between self and others and implements imitation behavior (The Best Paper of IEA·EAIE2005), *Innovations in Applied Artificial Intelligence, 18th International Conference on Industrial and Engineering Applications of Artificial Intelligence and Expert Systems*, IEA/AIE 2005, 101–110, 2005.

Takeno, J. Robot vision technology for mobile robots, *Journal of Information Processing Society of Japan*, 44(SIG 17(CVIM 8)), 24–36, 2003.

Takeno J, *The Self Aware Robot*, HRI-Press, August 2005.

Takeno J, MoNAD Structure and the self-awareness, *BICA2011*, Washington DC, 2011.

Takeno J, Inaba K, Suzuki T, Experiments and examination of mirror image cognition using a small robot, *The 6th IEEE International Symposium on Computational Intelligence in Robotics and Automation* (CIRA 2005), IEEE Catalog: 05EX1153C, ISBN: 0-7803-9356-2, 493–498, 2005.

Takeno, J, Mizuguchi, N, Sorimachi, K Realization of a 3D vision mobile robot that can avoid collision with moving obstacles, *Robotersysteme*, 8, 1–12, 1992.

Takeno, J, Rembold, U, Stereovision system for autonomous mobile robots, *Host Journal for the Intelligent Autonomous System Society, Robotics and Autonomous Systems*, 18, 355–363, 1996.

Tani J, On the dynamics of robot exploration learning, *Cognitive Systems Research*, 3(3), 459–470, 2002.

Turing AM, On computable numbers, with an application to the Entscheidungsproblem. *Proceedings of the London Mathematics Society (Series 2)*, 1936, 42, 230–265.

van Nes SI, Faber CG, *et al.*, Revising two-point discrimination assessment in normal aging and in patients with polyneuropathies, *Journal of Neurology, Neurosurgery, and Psychiatry*, 79, 832–834, 2008.

Watson, JB, *Behaviorism*, Norton, 1930.

Wehr, G, *An Illustrated Biography of C. G. Jung*, Rene Coeckelberghs Verlag AG, 1989.

Wiener, N, *I Am a Mathematician*, Doubleday & Company, Inc, 1956.

Wittgenstein, L, *Philosophical Investigations*, Basil Blackwell Ltd, 1953.

Wundt, W, Zur Kritik tachistosckopisher Versuhe, *Philosophische Studien*, 15, 287–317, 1899.

Web Sites

http://www.rs.cs.meiji.ac.jp/Takeno_Archive/DiscoveryNewsAwareRobot 211205.pdf.

http://dsc.discovery.com/news/briefs/20051219/awarerobot_tec_print. html.

http://www.rs.cs.meiji.ac.jp/Takeno_Archive.html.

http://www.consciousness.it/cai/online_papers/haikonen.pdf.

Index